MASTERING SERIAL COMMUNICATIONS

MASTERING SERIAL COMMUNICATIONS

SERIAL

COMMUNICATIONS

Peter W. Gofton

SYBEX® SAN FRANCISCO • PARIS • DÜSSELDORF • LONDON

Cover design by Thomas Ingalls + Associates
Book design by Joe Roter

To Mickey

TABLE OF CONTENTS

PART II: THE IBM PC AND PC DOS

PART III: SERIAL COMMUNICATIONS PROGRAMMING

ACKNOWLEDGMENTS

I would like to thank Reid Larson of DPEX for initiating me in the black art of communications, Mark Longwood for explaining the mysteries of IBM mainframe communications, Ward Christensen and Frank da Cruz for help with XMODEM and Kermit respectively, and many friends on the CompuServe Special Interest Groups for on-line help. Any errors, of course, remain my responsibility.

I would like to thank my agent Bill Gladstone, and Dr. Rudy Langer of SYBEX for trusting a new author with such an important subject, my editor Bonnie Gruen for patience in dealing with my ignorance on so many aspects of writing, and Elna Tymes for helping me with the proofreading. Thanks to Inmac and Hayes Microcomputer Products Inc. for supplying photographs.

Thanks too to the people "behind the scenes" at SYBEX who worked on this book: Ray Keefer and Joel Kroman, technical reviewers; Olivia Shinomoto and David Clark, word processors; Brenda Walker, typesetter; and Aidan Wylde, proofreader.

INTRODUCTION

*M*icrocomputer communications is an increasingly important subject. In business, electronic mail is already largely replacing telephone and mail communications within several large companies, electronic databases are being used more and more, and there is a large movement towards connecting micros within companies to the corporate mainframes. At home, interactive communications networks are becoming extremely popular, and electronic bulletin boards are springing up all over the country.

Nevertheless, information about the techniques involved is not easy to come by. Books about microcomputer communications tend to be concerned with how to *use* CompuServe, or how to *choose* communications software, rather than how to *write* communications software. Technical books on MS-DOS and programmers' guides to personal computers tend to give very little information about communications.

An otherwise excellent book *Programmer's Guide to the IBM PC* by Peter Norton (Microsoft Press, 1985) devotes four pages out of 412 to communications. *The Complete Handbook of Personal Computer Communications* by Alfred Glossbrenner (St. Martin's Press, 1985) has virtually no overlap with my book, despite its title.

The difficulty I have had in tracking down some of this material indicates, I think, the usefulness of this book. I believe that it is the first time that the XMODEM and Kermit protocols have been set out in a book; in each case I had to refer to the original developers for information.

Part One of the book is about serial communications in general. It includes information about hardware interfacing, software protocols, and modems. The information it covers applies to whatever machine you may have.

Part Two deals with the IBM PC. Its four chapters cover the subject in increasing detail, going from the user's point of view to detailed systems-level programming. Despite the primary focus of this section, many of the principles discussed will also be relevant to owners of machines other than the IBM PC.

Part Three gives actual programming examples in three languages: BASIC, C, and assembly language. These chapters assume that you will choose to read one of them and accordingly there is a certain amount of repetition among them. The final chapter, for advanced programmers only, describes an interrupt service routine written in C and assembly language.

Two last points:

1. Because the distinction between mainframes and minicomputers has become so blurred, I have used the word *mainframes* to cover *minicomputers* or *mainframes*.

2. All numbers are decimal unless otherwise stated; hexadecimal numbers are followed by the letter H.

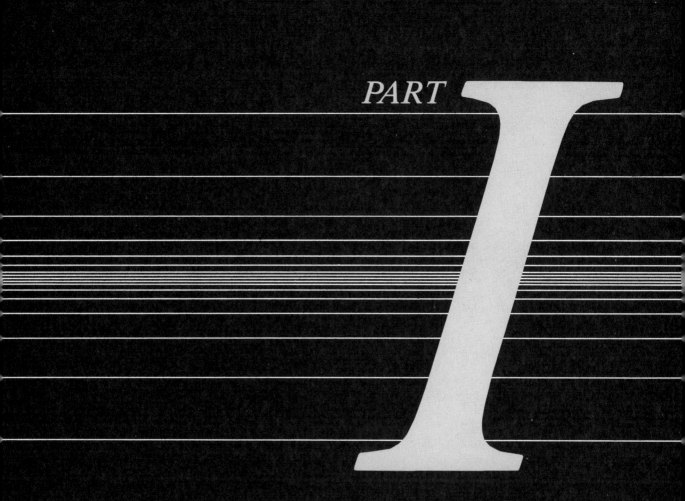

PART

I

OVERVIEW OF SERIAL COMMUNICATIONS

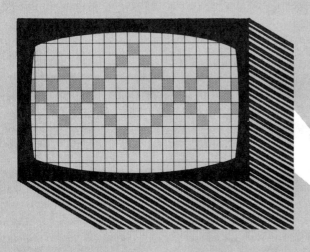

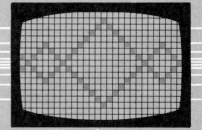

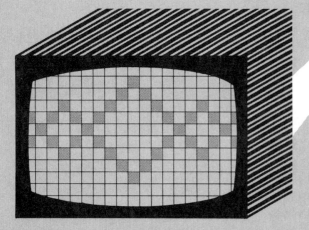

CHAPTER *1*

Hardware Interfacing

INTRODUCTION

*I*n order for two devices to be able to communicate, they must be connected in such a way that electrical signals transmitted by one are received by the other.

Communication can be achieved either directly, with wires connecting the two devices, or indirectly, with an intervening medium. This medium is most often the public telephone system, in which case *modems* (MOdulator DEModulators) are used to convert the signals at one end into signals suitable for transmission along telephone wires, and then convert them back at the other end. Other media, such as fiber optics cables and radio transmission, can also be used. These devices are connected to computers in much the same way as the more conventional serial devices; therefore, the principles described here will apply.

This chapter discusses the direct connection between two devices, including the cables and sockets required, and the commonly used standards that determine which wires are used for which purposes.

PLUGS AND SOCKETS

*T*here are several different types of plugs and sockets for connecting cables to serial devices. The 25-pin and 9-pin *D-type connectors* are the most common (sometimes referred to as DB-25 and DB-9), although there are other types in use, such as the circular DIN connectors used in some Apple computers. Some typical D-type connectors are shown in Figure 1.1.

D-type connectors (so named because they are shaped somewhat like the letter D) contain a certain number of pins or sockets. Those with pins are *male* connectors and those with sockets are *female* connectors. Each pin or socket has a number, which is generally printed on the connector adjacent to each pin. The pin connections on some common connectors are shown in Appendix A.

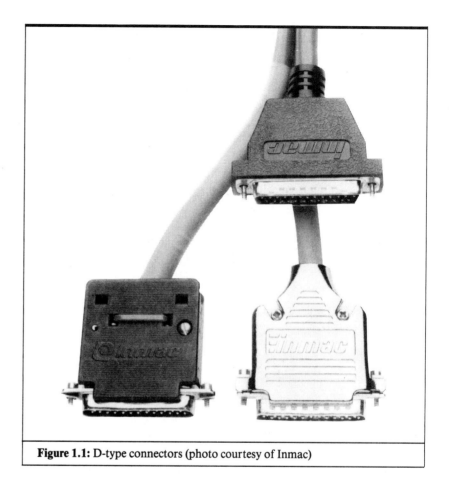

Figure 1.1: D-type connectors (photo courtesy of Inmac)

THE RS-232-C STANDARD

*I*n order to make equipment from different manufacturers compatible, various standards have been designed. The most widely used is RS-232-C, published in 1969 by the Electronic Industries Association. The RS-232-C standard was originally drawn up to specify the connections between terminals and modems. It specifies the electrical characteristics of circuits between the two devices and gives names and numbers to the wires necessary for joining them. The circuit names allotted by RS-232-C (AA, AB, etc.) are hard to remember, and have been replaced in actual practice by abbreviations.

Let's look at line 2 as an example. This line is officially known as BA but more commonly as TXD (Transmitted Data). According to the RS-232-C standard, line 2 carries data from the terminal to the modem. For this to operate correctly, the terminal must produce output at line 2 and the modem must receive data on line 2. Therefore, line 2 is a transmitting line for some devices and a receiving line for others. A direct connection from pin 2 to pin 2 (and so on with the other pins) can only be made when one device transmits on line 2 and the other receives on line 2. Otherwise, both would be trying to transmit on the same line, and successful transmission of data would not be possible.

DTE AND DCE

*T*o prevent devices from attempting to talk to each other along the same wires, devices are divided into two types. Devices such as terminals which use pin 2 for output are known as DTE (Data Terminal Equipment). Devices such as modems which use pin 2 for input are known as DCE (Data Communication Equipment).

According to RS-232-C, DTE devices should have male connectors and DCE devices should have female connectors. However, manufacturers do not always comply with this rule and it is not always immediately obvious whether a given device is DTE or DCE.

When you know that one device is DTE and the other is DCE, you can, at least in theory, connect them easily by connecting pin 2 to pin 2 and so on with the other pins. This is known as a straight connection. Not all manufacturers comply with the standard, unfortunately, and, as a result, several problems result. I will discuss these problems, and how to deal with the situation where both devices are DTE or both DCE, later in this chapter. For the time being, let's assume that one device is DTE and the other DCE, and that each is supplying the signals required by the other on the corresponding pins.

ONE-WAY COMMUNICATION

*T*here are three main circuits that are used for communication: line 2 (data from DTE to DCE), line 3 (data from DCE to DTE), and line 7 (signal ground). Signal ground serves as a common reference point from which the polarity and voltage of the other lines can be determined. In the simplest case, where only one device transmits and one receives, only two wires need to be connected: line 2 or 3 and line 7. Figure 1.2 illustrates the simplest form of communication.

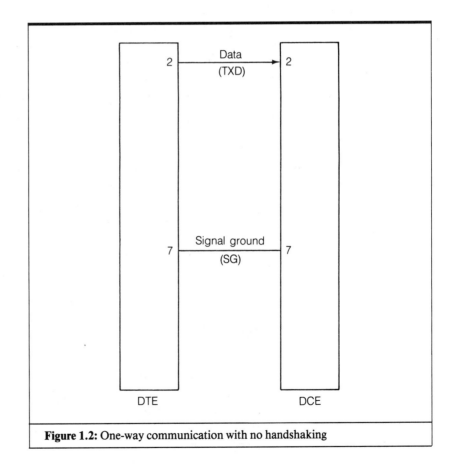

Figure 1.2: One-way communication with no handshaking

HARDWARE HANDSHAKING

In many cases, it is necessary for the transmitting device to know whether the receiving device is ready to receive information. You might, for example, be sending data to a printer, and the speed of communication may be faster than the speed of the printer. The printer will have to be able to stop the computer from sending any more characters until it has printed out the ones it has already received. Similarly, you may be sending data from one computer to another, and the second computer cannot process the data as fast as it is coming in.

In both of these cases information must be sent back from the receiving device to the transmitting device to indicate whether it is ready or not. This information is known as *flow control* or *handshaking.*

There are two types of handshaking: hardware and software. Both involve signals coming back from the receiving device to the transmitting device. With hardware handshaking, the receiving device sends a positive voltage along a dedicated handshaking circuit as long as it is ready to receive. When the transmitting computer receives a negative voltage, it knows to stop sending data. With software handshaking, described in Chapter 3, the handshaking signals consist of special characters transmitted along the data circuits rather than handshaking circuits.

To incorporate hardware handshaking, at least one additional wire connection must be made to carry the signal. This brings the total number of wires to three: transmitted data, common, and handshaking.

DTE TO DCE

When a DTE device is transmitting to a DCE device, line 2 is used for the data and line 7, as usual, carries the signal ground. A DCE device normally controls handshaking transmissions from a DTE device on wire 6, known as DSR or Data Set Ready. If the printer is DCE and the computer DTE, pin 6 on the printer should be connected to pin 6 on the computer, and the printer will maintain a positive voltage on pin 6 as long as it is able to receive data. When it wishes the computer to stop sending data, it will drop the voltage on pin 6 to a negative state.

Often a second handshaking line, circuit 5, Clear to Send (CTS) is also used by a DCE device to control transmissions from a DTE device. Where two handshaking wires are used, the DTE device must be designed to transmit only when both lines are high, or positive. Sometimes the lines have different meanings. For example, one might tell the transmitting device to stop printing until a certain amount has been printed, and the other might indicate that the printer is out of paper. However, these meanings are not standardized. Since many computers are programmed not to transmit unless both handshaking lines are high, even printers that do not allocate a special meaning to the second line should at least maintain a positive voltage on it; not all of them do, however, and sometimes the second signal must be faked by joining it to the first.

Figure 1.3 illustrates one-way communication from a DTE device to a DCE device with two handshaking wires.

DCE TO DTE

In order for a DCE device to talk to a DTE device, line 3 must be used for the data transmissions, and if handshaking is required, line 20 must be used to send handshaking from a DTE device to a DCE device. Line 20 is known as DTR or Data Terminal Ready. The secondary handshaking wire, not always used, is line 4, Request to Send (RQS). Figure 1.4 illustrates DCE to DTE communications with handshaking.

TWO-WAY COMMUNICATIONS

*D*ata are often transferred in two directions. This usually occurs when two computers communicate with each other, but also occurs in communication between other devices when software handshaking is being used. The minimum number of wires necessary in two-way communication is three: transmitted

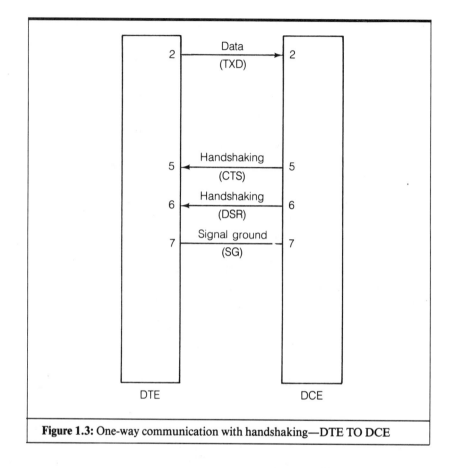

Figure 1.3: One-way communication with handshaking—DTE TO DCE

data in each direction and signal ground. The addition of one handshaking line in each direction brings the total to five, as shown in Figure 1.5.

When secondary handshaking lines are added in each direction, the total comes to seven. Two additional lines are often added that enable a modem to give more information to a computer or terminal. CD (Carrier Detect) is connected to pin 8, and is used to indicate the presence of a carrier signal. RI (Ring Indicator) is connected to pin 22, and indicates that the modem is being called by a remote device and would be ringing if it were a telephone. The total number of circuits now comes to nine, and these are shown in Figure 1.6. Although many other circuits are defined by RS-232-C, these

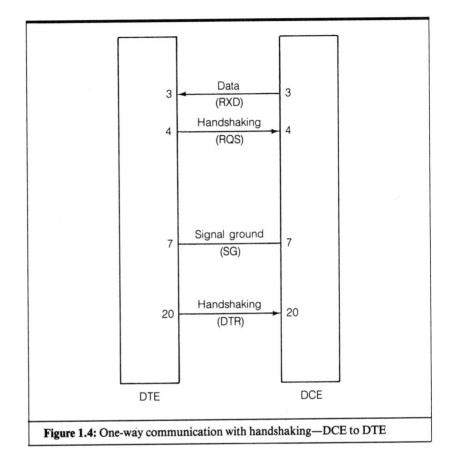

Figure 1.4: One-way communication with handshaking—DCE to DTE

nine are the most common and are the only ones normally connected to microcomputers. This is why microcomputers tend to use 9-pin connectors now rather than the 25-pin type that would be necessary to carry the full set of RS-232-C circuits.

NULL MODEMS

As I stated earlier, RS-232-C was originally devised to specify connections between terminals, which are DTE, and modems, which are DCE. However, it has been extended to apply

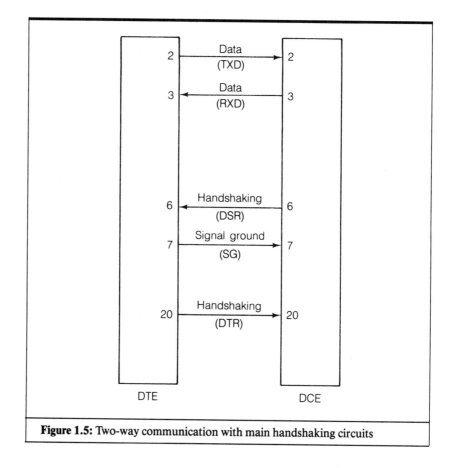

Figure 1.5: Two-way communication with main handshaking circuits

to connections between many other kinds of devices that were not officially specified as being either DTE or DCE, such as micro-computers and printers.

Since there is no standard that indicates whether certain devices should be DTE or DCE, often you will have to connect two DTE devices or two DCE devices. In this case, you must connect pin 2 on the first device to pin 3 on the second, and pin 3 on the first device to pin 2 on the second. The handshaking wires must be crossed in the same way.

You can cross the wires by either connecting the devices with a cable in which the wires are crossed over, or by buying a special connector that connects both devices and performs the necessary

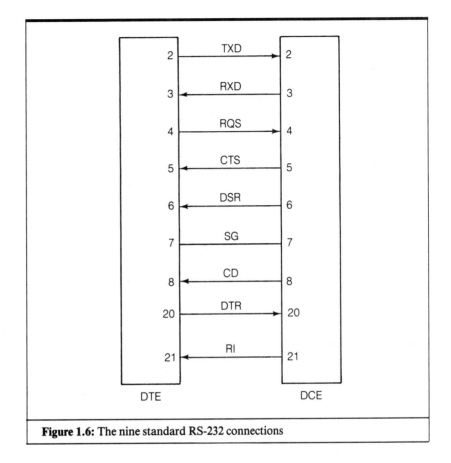

Figure 1.6: The nine standard RS-232 connections

crossing internally. In either case, the intervening cable or connector is called a *null modem* because it takes the place of two modems (each DCE) that enable two DTE devices to talk to each other—DTE : DCE : DCE : DTE. The wiring for a null modem is shown in Appendix A.

ELECTRICAL SIGNALS

*T*he RS-232-C standard lays down the characteristics of the electrical signals used in direct serial connections. There are only two states permitted: *SPACE*, corresponding to binary zero, where a positive voltage exists, and *MARK*, corresponding to binary one, where a negative voltage exists.

On data lines (i.e., on lines 2 and 3), a positive voltage (SPACE) corresponds to a logical zero, and a negative voltage (MARK) corresponds to a logical 1. On handshaking lines (e.g., DTR, DSR), a positive voltage (SPACE) indicates that the line is ON, meaning "go ahead and send." A negative voltage (MARK) means "stop."

Positive voltages (the SPACE state) are between $+5$ and $+15$ volts for outputs, and between $+3$ and $+15$ volts for inputs. The difference allows for voltage loss arising from cable length. Negative voltages (the MARK state) are specified as being between -5 and -15 volts for outputs, and between -3 and -15 volts for inputs.

Note that if too long a cable is used, the voltage levels fall outside the permitted boundaries. In addition, a build up of capacitance affects the quality of the signal by smoothing out the transitions from positive to negative voltage. RS-232-C is not intended to be used for long distances, and 50 feet is generally considered the maximum distance using normal cable at usual transmission rates. If devices are too far apart, a modem or some other method of communication is necessary.

RS-449

A newer standard than RS-232-C, RS-449 tackles the same problems but allows for greater transmission speeds and a reduction in electrical crosstalk. RS-449 specifies a 37-pin connector and, in case that is not enough, offers an additional 9-pin connector. If you have ever soldered as many 25-pin plugs as I have (25 pins within the space of 1 ½ inches), you will not be overly enthusiastic about RS-449, which has not been widely adopted. RS-449 covers the mechanical specifications and circuit descriptions, but not the electrical characteristics. It is intended to be used in conjunction with RS-422-A and RS-423-A, which describe the electrical characteristics of balanced and unbalanced circuits respectively. Balanced circuits, used in higher speed applications or where transmission problems are likely, use two wires,

A and B. The MARK and SPACE conditions are signaled by changes in the polarity of the two wires by reference to each other, instead of a single wire changing in polarity by reference to a single common or signal ground.

It is possible to interconnect RS-449 and RS-232-C devices. The Electronic Industries Association (EIA) publishes a document entitled "Application Notes on Interconnection Between Interface Circuits Using RS-449 and RS-232-C," or Industrial Electronics Bulletin No. 12. However, this document only applies to RS-423 (unbalanced) circuits.

Incidentally, the EIA has been trying to encourage the use of the designations EIA-XXX rather than RS-XXX. You will see the standards referred to both ways but in this book I will refer to them as RS-XXX.

THE APPLE MACINTOSH

The serial ports on the Apple Macintosh use balanced circuits and Apple claims compliance with RS-422. However, its 9-pin DB connectors bear no relationship to the pin connections specified in RS-449. Since the Macintosh incorporates RS-422 (balanced) circuits, it is not theoretically possible to interface it with RS-232-C circuits because RS-422 does not have a common signal ground by reference to which both the transmitted data line and the received data line vary. Nevertheless, the Macintosh has been designed in such a way as to make this possible. The ImageWriter printer often used with the Macintosh is a DTE RS-232-C device, and the cable used for connecting the Macintosh to the Image-Writer has a DB-25 connector, which can be used for interfacing to other serial devices.

TROUBLE SHOOTING

*T*he following are some notes to help with problems you may encounter while connecting serial devices together.

IS A NULL MODEM NEEDED?

As I mentioned above, a null modem is needed whenever you have to connect two devices that are both DTE or both DCE. In some cases, however, you may not know whether a device that you are trying to connect is DTE or DCE (newer devices such as mice and light pens often cause this problem). Therefore, since there is always the possibility that the two devices might be the same, if you cannot transmit with a straight connection, you should try using a null modem.

HANDSHAKING PROBLEMS

If a printer does not respond to the transmitting device, it might be that the printer requires two handshaking lines to be high whereas the computer is only providing one. This is often the case with the IBM PC, which is capable of providing two lines but, unless specially programmed, provides only one. It is often possible to fake the second signal by connecting the subsidiary handshaking wire to the main one at the printer end of the cable.

If the microcomputer is not providing any handshaking signals at all, and the printer insists on receiving one or two, even more faking can be carried out by feeding the printer's own outgoing handshaking signal back into the printer as an incoming signal.

If a computer is not sending when it should be, it is probably because it is expecting a handshaking signal which it is not receiving. If a printer is only setting one handshaking line high, try connecting the other handshaking line to it at the computer end.

USING BREAK-OUT BOXES

If you do a lot of interfacing, I strongly recommend investing in a *break-out box*. This small device has two D-type connectors that can be inserted between two serial devices. It has a light for each pin that lights up when there is a signal on its respective wire. The break-out box, therefore, allows you to see when data are being transmitted, and along which line, and also which handshaking lines are carrying a positive voltage. You can switch wires in and out and join wires by inserting jumpers into sockets.

By using a break-out box you can experiment with changing connections without having to solder and desolder wire. Having established which connections are correct, you can then make up the appropriate cable. A typical break-out box is shown in the photograph at Figure 1.7.

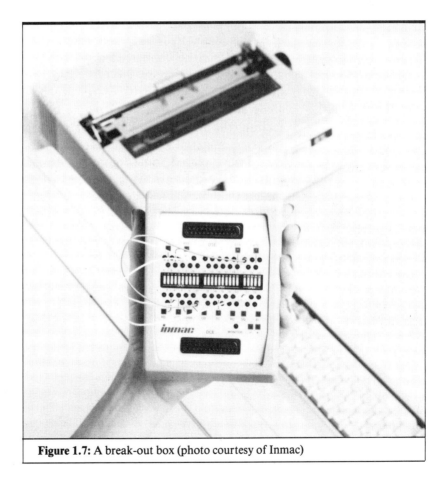

Figure 1.7: A break-out box (photo courtesy of Inmac)

OTHER PROBLEMS

If your receiving device is receiving garbage, the problem is probably one of character transmission. See Chapter 2 for more information.

For examples of interfacing various devices, and more information about RS-232, refer to *The RS-232 Solution* mentioned in the bibliography.

SUMMARY

*I*n this chapter I have described the main ways that serial devices can be connected together. I have described the plugs and sockets, the main circuits, and the voltage levels. I have introduced the concept of hardware handshaking, and outlined some suggested problem-solving techniques. You may find after reading the book that you want to refer back to some of the reference material I have included on this subject. For easy reference I have placed some figures in Appendix A that illustrate the pins and connectors of a number of common connectors. In the next chapter, I will describe how data are converted into suitable form for transmission.

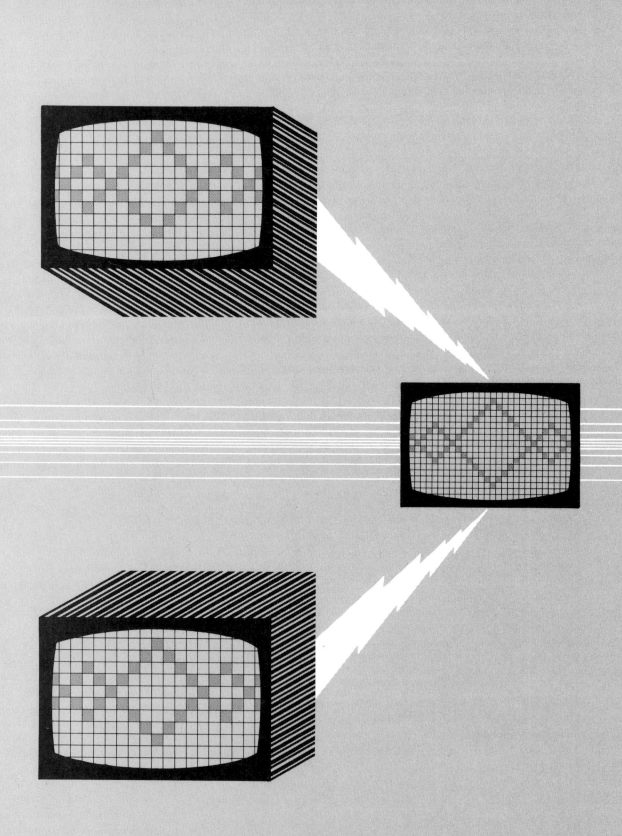

CHAPTER 2

Character Transmission

INTRODUCTION

*I*n the last chapter, we described the hardware connections between two devices. In this chapter we will explain how individual characters are encoded and sent along the wires. The principles discussed here apply both to signals sent along telephone wires between modems, and signals sent along cables between computers.

DATA FORMAT WITHIN THE COMPUTER

*I*n order to understand the transmission of data, we must first understand how it is stored within the computer.

BITS AND BYTES

In decimal notation, there are ten digits—zero to nine. Adding a zero to the end multiplies a number by ten. In *binary notation,* there are only two digits—zero and one. Adding a zero to the end of a number multiplies the number by two. For example, the series 0 1 2 3 4 5 in decimal would be 0 1 10 11 100 101 in binary.

Each binary digit of zero or one is known as a *bit.* Eight bits make up a byte. Accordingly, the values for a byte can range from 00000000 to 11111111, or 0 to 255 in decimal.

By convention the rightmost bit in a byte is referred to as *bit zero.* The leftmost bit is known as *bit seven.* Bit zero is also known as the *least significant bit,* and bit seven as the *most significant bit.* Figure 2.1 shows the number 35 in binary.

Almost all computers work in binary because it is easy to code ones and zeros as positive and negative voltages. In most computers, the smallest unit of storage that can be referred to by means of a memory address is the byte. Therefore, when information is stored and manipulated in a computer it is normally converted into a sequence of bytes.

Bit number	7	6	5	4	3	2	1	0
Value if set	128	64	32	16	8	4	2	1
Setting	0	0	1	0	0	0	1	1
Value as set	0	0	32	0	0	0	2	1

Figure 2.1: Number 35 in binary

CODING TEXT

When text (alphabetical characters, punctuation marks, etc.) is stored in a computer, each different character is represented by a different number. These numbers normally range from zero through 127, or from zero through 255. Since a byte can have a value from zero through 255, it is natural to allocate one byte to each letter or punctuation mark in text data.

There are two different conventions for mapping characters to numbers: EBCDIC (Extended Binary Coded Decimal Interchange Code), which is used in IBM computers other than the IBM PC series, and ASCII (American Standard Code for Information Interchange), which is used in most other computers. We will only be dealing with the ASCII method in this book.

The official ASCII table gives a number between 32 and 126 to all the commonly used characters (upper- and lowercase), numbers, punctuation marks, and other symbols. The numbers from zero to 31 and 127 have special meanings such as carriage return, line feed, and other *non-displayable characters.*

For example, uppercase A is stored as decimal 65. In binary this is 01000001. A comma is stored as decimal 44, which is 00101100 in binary.

Since the number 127 in binary uses seven bits, all the characters represented by zero through 127 can be stored in one byte, leaving one extra bit. Because we name the bits in a byte zero through seven, we can see that the ASCII code uses only bits zero through six. Bit seven is spare.

Many computers use the full eight bits of each byte for coding letters, giving a total of 256 different combinations. The first 128 follow the ASCII mapping, and the remainder are used for foreign characters, mathematical symbols, graphics characters, and so on, as the designer wishes. Unfortunately, there is no standard for these *extended characters,* which tend to have different meanings on different computers.

SPECIAL ASCII CHARACTERS

The first 32 codes in the ASCII table do not represent printable characters, but have special meanings. Many of them were specifically designed to aid communications. Some of them exist for historical reasons.

Codes 1 through 26 are also referred to as Ctrl-A through Ctrl-Z, and can normally be generated on a computer keyboard by holding down the key marked *Ctrl* and pressing the appropriate alphabetical key at the same time. Some of the codes can also be achieved by pressing dedicated keys, such as Tab for code 9 or Return for code 13.

The special ASCII codes are listed below.

#	Character	Description
0	NULL	A method of deliberately causing a delay. At one time with slow printers it was necessary to send nulls after each carriage return to allow the printer carriage to return to the left-hand edge of the page.
1	SOH	Start of heading: indicates that the following text is part of a title.
2	STX	Start of text: indicates the start of the actual text of the message.
3	ETX	End of text.
4	EOT	End of transmission.

#	Character	Description
5	ENQ	Enquiry: normally used as part of a software handshaking sequence asking the receiving computer to acknowledge receipt of the message.
6	ACK	Acknowledges receipt of a message.
7	BEL	Rings the terminal bell or equivalent.
8	BS	Backspace.
9	HT	Horizontal tab.
10	LF	Line feed: causes a skip to the same position one line below.
11	VT	Vertical tab.
12	FF	Form feed: advances to a new page.
13	CR	Carriage return: moves to the beginning of the line. Sometimes also causes a line feed, but this varies.
14	SO	Shift out: marks the start of a special control code sequence. Esc is often used instead now (see below).
15	SI	Switch in: marks the end of a control code sequence initiated by SO.
16	DLE	Data link escape: Similar to Esc.
17	DC1	Device control 1 to 4: four spare codes to be used as desired; often used in software handshaking.
18	DC2	
19	DC3	
20	DC4	
21	NAK	Negative acknowledgement: indicates that a transmission was not received correctly. For example, a parity error may have been detected.
22	SYN	Synchronous idle: similar to a NULL, but used in synchronous communication to keep two devices synchronized

#	Character	Description
		in between transmission. Synchronous communications are described later in this chapter.
23	ETB	End of transmission block: used where transmissions are divided into blocks for error checking purposes.
24	CAN	Cancel: disregard the data sent.
25	EM	End of medium: indicates approaching end of a paper tape.
26	SUB	Substitute: corrects an erroneously sent character. Also used in practice to indicate end of transmission.
27	Esc	Escape: indicates the start of a sequence of characters with special meaning to the recipient.
28	FS	File separator ⎫
29	GS	Group separator ⎬ Mark boundaries
30	RS	Record separator ⎬ between text segments.
31	US	Unit separator ⎭

CODING NONTEXTUAL MATERIAL

Of course, not all the material stored in a computer is in text form. Program instructions, numeric data, and graphic images, for example, are not stored in ASCII form.

These types of data are normally coded in such a way as to use all 256 possible values of a byte. Numbers are stored in binary form and can extend over several bytes. Program instructions often consist of one or two bytes. We often refer to this type of material, in a communications context, as *binary data,* even though text is also stored in binary form.

Since the bytes holding nontextual data can be of any value, at times they will correspond to values that have special meanings in the ASCII table. This can cause complications if you are transmitting data and your receiving device happens to interpret a nontextual byte to mean the end of the message. In this case, the data cannot be sent in their *raw* form because a byte in the middle of the message might accidentally correspond to the end of message symbol and the receiving device would stop listening.

Accordingly, certain protocols have been designed to cope with this problem; some are described in Chapters 6 to 9.

CONVERSION TO SERIAL FORM

*A*lmost all computers store and manipulate their data in parallel. This means that when a byte is sent from one part of the computer to another it is not sent one bit at a time but several bits at a time over a number of wires running in parallel. The number of bits sent at a time varies from machine to machine but is normally eight or a multiple of eight. Therefore, a computer can work with at least one byte and often two or more bytes at once.

Since communication from a computer to another device is done serially, meaning that data are sent one bit at a time, a serial interface must take bytes that are received in parallel, and send out the individual bits separately.

As we have seen, the data lines in serial communications can only be in either MARK or SPACE condition, which in direct connection equates to negative or positive voltages, respectively. Any transmitted data must first be translated into a sequence of MARKs and SPACEs. For the purposes of this translation, a MARK represents a one, and a SPACE represents a zero.

SYNCHRONOUS AND ASYNCHRONOUS COMMUNICATIONS

*O*nce the data are converted to serial form, there are two ways that they can be communicated: *synchronously* or *asynchronously.*

When data are being transmitted by someone typing at a keyboard, they are almost always sent and received asynchronously. A person typing at a keyboard cannot type at a continuous, even pace; therefore, when the computer receives the letters, there are differing gaps between each character. If the individual letters are being serially transmitted as they are typed, the irregular gaps between the characters make it impossible for a receiving device, after receiving one character, to tell exactly when the next one will arrive. Because of this lack of continuity, it is necessary to place extra bits before and after each character to indicate to the receiving device the beginning and end of the character. These extra bits are known as *start bits* and *stop bits*. In addition, another bit, known as the *parity bit,* is often added to enable errors to be detected (described below). This method is known as *asynchronous communication.*

When characters are sent in a block at machine speed, they can be spaced out regularly. It is no longer necessary for each character to have start and stop bits, because once the first character has been received the receiving device can predict exactly when the following characters will arrive. In other words, it can synchronize itself with the transmitting computer. This method is known as *synchronous communication.*

There is considerable overhead involved in adding the extra bits necessary for asynchronous communication. A transmission can take about 20 percent longer than when using synchronous methods. Of course, when the whole process is being governed by human typing speed, there would not be any savings. But when large blocks of data are sent over expensive telephone lines, synchronous communications are much more efficient.

Nevertheless, asynchronous adapters and modems are simpler and cheaper than synchronous equivalents. For that reason, and because most devices with which microcomputers communicate are themselves asynchronous, very few microcomputer owners use synchronous communications. Therefore, the remainder of this chapter will concentrate on asynchronous communication.

FRAMING

*I*n the case of asynchronous serial communications, the bits representing one byte, which are known as the *data bits,* are preceded and followed by start, stop, and parity bits, which are described fully in this section. This process is known as *framing*.

The number of bits representing one character varies according to the communications protocol in use. It is known as the number of data bits, or the *word length*. It is normally either seven or eight bits. Each character is sent in a group consisting of a start bit, the character (data bits), an optional parity bit, and one or more stop bits. For the sake of clarity, I will refer to each group consisting of one character and its associated bits as a *frame,* in order to avoid the confusion that can occur when the word *character* refers sometimes to the data bits and sometimes to the full group with start, stop, and parity bits. Two examples of transmitted frames are shown in Figure 2.2.

Transmitting the letter A with seven data bits, one stop bit, and even parity:

	Start bit	Data bits		Parity bit	Stop bits
Logical	0	1 0 0 0 0 0 1		0	1
SPACE or MARK	S	M S S S S S M		S	M
Voltage (+ or −)	+	− + + + + + −		+	−
Time →					

Transmitting the letter A with eight data bits, two stop bits, and odd parity:

	Start bit	Data bits		Parity bit	Stop bits
Logical	0	1 0 0 0 0 0 1 0		1	1 1
SPACE or MARK	S	M S S S S S M S		M	M M
Voltage (+ or −)	+	− + + + + + − +		−	− −
Time →					

Figure 2.2: Transmitting the letter A

START BITS

A start bit is always added at the beginning of a frame to alert the receiving device that data are arriving and to synchronize the mechanism that separates out the individual bits. A start bit is a SPACE, or binary 0.

With direct connection, a SPACE or 0 is transmitted as a positive voltage. The voltage between frames is negative. Accordingly, at the start of each frame, the voltage changes from negative to positive.

DATA BITS

The serial communications standards, called *protocols,* allow for the transmission of different lengths of characters, or words. When communications software asks you to select word length, it is asking whether you want to send seven- or eight-bit characters (sometimes other lengths are used but this is rare). If all the data to be transmitted are in ASCII form, seven-bit words are sufficient. Remember that the ASCII table only assigns numbers from 0 to 127, all of which can be represented in seven bits.

If non-ASCII data are to be transmitted (for example, text using extended character sets or binary data), all eight bits of each byte are needed and you cannot use a seven-bit protocol unless the data are first converted into seven bit format. This is discussed more fully in Chapter 7.

Data bits are transmitted least significant bit first as opposed to the way they are set out in writing. Thus the letter A, which is ASCII 65 decimal, would normally be printed as 01000001 in binary, but is transmitted 10000010 as eight bits, or 1000001 as seven bits.

PARITY BIT

Parity checking is a method of testing whether the transmission is being received correctly. The sending device adds a parity bit, which is calculated according to the contents of the data bits. The receiving device checks that the parity bit does indeed bear the correct relationship to the other bits. If it does not, something must have gone wrong during the transmission. Parity can be computed in any of the ways discussed below.

Even Parity

Even parity means that adding the data bits and the parity bit yields an even number. For example, the letter A in binary is 01000001. When you add up the bits you get 2—an even number. Since the total of the bits must be even, the parity bit must be zero.

If the letter A is received with the parity bit set (i.e. equal to 1), an error must have occurred during transmission.

Odd Parity

Odd parity means that the total of the data bits plus the parity bit yields an odd number. So, again using the letter A, the parity bit would have to be set to 1, to bring the total of the bits to 3—an odd number.

No Parity

A parity bit is not always used, and it is often ignored by the receiving device even when it is used. It all depends on how the two devices have been programmed. *No parity* means there is no parity bit.

SPACE Parity

Sometimes a parity bit will be used, but always set to zero. This provides some error checking, since with seriously garbled transmissions the parity bit would sometimes be set indicating an error of some sort.

SPACE, or zero, parity can also be used to transmit seven-bit characters to a device that is expecting eight-bit characters. The receiving device interprets the parity bit as the last bit of the data word. However, when all the characters sent are standard ASCII characters, the last bit is never used. SPACE parity is sometimes referred to as *bit trimming*.

MARK Parity

MARK parity works the same way as SPACE parity except that the parity bit is always set to 1. It is sometimes referred to as *bit forcing*.

STOP BITS

At the end of each frame, stop bits are sent. There can be one, one and a half, or two stop bits. *One and a half bits* means that the length of the bit is greater than that of a normal bit. The stop bits force a certain minimum gap between frames. They are sent as binary 1s, which, in direct connection, equal a negative voltage.

There is always at least one stop bit. This ensures that there is a negative voltage for at least some period of time between two frames so that the next frame can be recognized by its positive start bit. More than one stop bit is generally used when the receiving device requires extra time before it can handle the next incoming character. Two stop bits are usually used at 110 baud which is the lowest transmission rate in general use. This is consistent with the requirements of older teletypewriter terminals, which use a low baud rate and require extra time to process characters.

BREAK

As explained, the data line between characters is normally in MARK condition (negative voltage, binary 1). It is only in the SPACE condition (positive voltage, binary 0) while a frame is being transmitted. If a character consists of all zeros, with eight data bits, and even parity, the SPACE condition remains for the start bit, the eight data bits, and the parity bit: i.e., ten bits. The

SPACE condition ends when it reaches the stop bit, which is negative. Even with each of these ten bits set as positive, at the rate of 150 bits per second, the SPACE condition would last only 1/15 second, or 66.67 milliseconds.

A longer SPACE condition than this, normally 100 to 600 milliseconds, is used as a special signal known as a *break*. The break is sometimes used as the mainframe's equivalent of Ctrl-C. It interrupts whatever program is currently running and returns the user to the operating system, or to some earlier level of a menu hierarchy within a program. Like Ctrl-C, it is useful for getting out of a program that has gone into an endless loop.

BAUD RATE

*T*he *baud rate* is the length of the shortest signal divided into one second. It is named for the French communications pioneer Baudot. *Bits per second* (bps) means the number of binary digits transmitted in one second. There is a difference between the two, but they are often confused. Probably 200,000 people would tell you they have 1200 baud modems, and not one of them actually has one! They are actually 1200 bps modems.

In direct RS-232 connections, a signal is in one of two states at any one time and the baud rate and the bps rate are the same. However, as we will see in Chapter 4, when a signal is transferred between modems it can be in one of several states. The signal length may be 1/600 seconds (600 baud) but, since more than two bits of information can be transmitted with each change of state, the bps rate will be higher than the baud rate.

It is important to note that both baud rate and bps refer to the rate at which the bits within a single frame are transmitted. The gaps between the frames can be of variable length, usually because the characters are being keyed in at a variety of speeds. Accordingly, neither baud rate nor bps refers to the rate at which information is actually being transferred.

When each new character is received, the receiving device is resynchronized. Therefore, a start bit is needed to signal the start of a new frame and to trigger whatever mechanism is used by the receiving device to read and interpret the bits that follow.

Bps rates are generally in the series 110, 150, 300, 600, 1200, 2400, 4800, 9600, and 19200. The most common rates for modem communications are 300 and 1200. 1200 is common for computer to printer communications, and 9600 for terminal to computer connections.

TROUBLE SHOOTING

When two devices are set up to communicate with each other, they must each agree on baud rate, word length, number of stop bits, and parity. If you find that nothing at all is being received, then the error probably lies in the physical connection: the data are being sent along the wrong line, there is a break in the line, or the correct handshaking signals are not being received. If garbage is being received, then the error probably lies in one of the areas discussed below.

BAUD RATE MISMATCH

If the two devices are set at different baud rates, the receiving device may attempt to interpret the data (unless it is programmed to report parity and framing errors). Typically, you will see that the number of received characters differs from the number sent.

WORD-LENGTH MISMATCH

If eight-bit words are being sent and the receiving device is expecting seven-bit words, you may not notice any difference in text transmissions, because often only the first seven bits are significant anyway. Because bit zero is sent first and bit seven is not used in genuine ASCII transmissions, its omission is not necessarily important. However, the receiving device may try to interpret the extra bit as a parity bit and report an error. Accordingly a *parity error* does not necessarily mean that the data have been damaged in transmission, it may indicate a word-length mismatch.

If seven-bit words are being sent, and the receiving device is expecting eight-bit words, the parity bit may be treated as the missing bit seven. Since the parity bit is generally 1 for half the characters and 0 for the other half, you will often find that the receiving device will display extended characters, such as graphics characters, in place of half the characters received.

PARITY ERROR

A *parity error,* strictly speaking, indicates that data have been damaged in transmission. However, it can also mean that the two devices have not been set to agree on parity (odd, even, or none) or on word length.

STOP BITS

There should be no problem if two stop bits are sent and only one is expected. The extra stop bit simply merges into the gap that is permitted between characters. However, sending one stop bit when two are required could cause a problem depending on the characteristics of the receiving device. It is not likely to be a problem with modern equipment.

FRAMING ERROR

A *framing error* indicates a mismatch in the number of bits and is usually reported when an expected stop bit is not received.

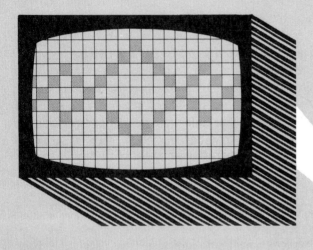

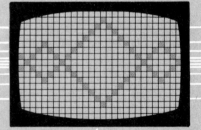

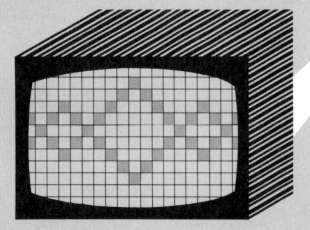

CHAPTER 3

Handshaking and Buffers

INTRODUCTION

*H*andshaking refers to methods with which a receiving device can control the flow of data from a transmitting device. Sometimes a printer cannot print characters as fast as it receives them from a sending computer. It must use handshaking to cause the computer to suspend transmission. Handshaking is also useful when the printer runs out of paper or when a computer sends data to another computer and the receiving computer cannot process the characters as fast as they are being received.

When you know that a receiving device can process received characters faster than the transmission rate, you can dispense with handshaking.

HARDWARE HANDSHAKING

*H*ardware handshaking, as discussed in Chapter 1, is the use of dedicated handshaking circuits to control the transmission of data. To summarize: DCE equipment normally uses DSR (Data Set Ready) as a main handshaking line to tell DTE that it is powered up and ready to control transmissions it is receiving. It can also use CTS (Clear to Send) as a subsidiary handshaking line. DTE equipment, on the other hand, uses DTR (Data Terminal Ready) as a main handshaking line to tell DCE that it is ready to receive, and RQS (Request to Send) as a subsidiary handshaking line. By convention, these handshaking wires carry a positive voltage when transmission is to be enabled, and a negative voltage when it is to be suspended.

For example, a serial printer configured as a DTE device will raise DTR to a positive voltage when it is ready to receive characters and lower DTR to a negative voltage when it wants to suspend the transmission. It can also use RQS as a subsidiary handshaking line. As explained in Chapter 1, if the computer is also a DTE device, a null modem must be used to transpose the signals. This means that DTR and RQS from the printer become

DSR and CTS at the computer. Many computers are programmed not to transmit unless both DSR and CTS are high.

The flowchart in Figure 3.1 illustrates a typical sequence for a computer sending data to a printer using hardware handshaking.

SOFTWARE HANDSHAKING

*W*hen handshaking signals are sent as data along the data wires (TXD and RXD, lines 2 and 3), instead of along dedicated handshaking circuits as in hardware handshaking, this is called *software handshaking.* This method is generally used where two computers are communicating (either directly or via a modem) and when two-way communication is possible.

Several standard protocols have been established to govern software handshaking, the most common of which is XON/XOFF.

XON/XOFF

Under this protocol, the receiving device sends ASCII character DC3 (19 decimal, 13H) to the transmitting device when it wants to stop the transmitting device from sending characters. It sends ASCII character DC1 (17 decimal, 11H) when it wants the transmissions to resume.

In normal practice, a buffer will be implemented. The DC3 character will be sent to the transmitter when the buffer is almost full and the DC1 character will be sent when the buffer is almost empty. (Buffers are described later in this chapter.)

The flowchart in Figure 3.2 illustrates a typical sequence for a computer sending data to a printer using XON/XOFF protocol.

ETX/ACK

In the end-of-transmission/acknowledge method, data are sent in fixed-length batches. After sending each batch, the transmitting

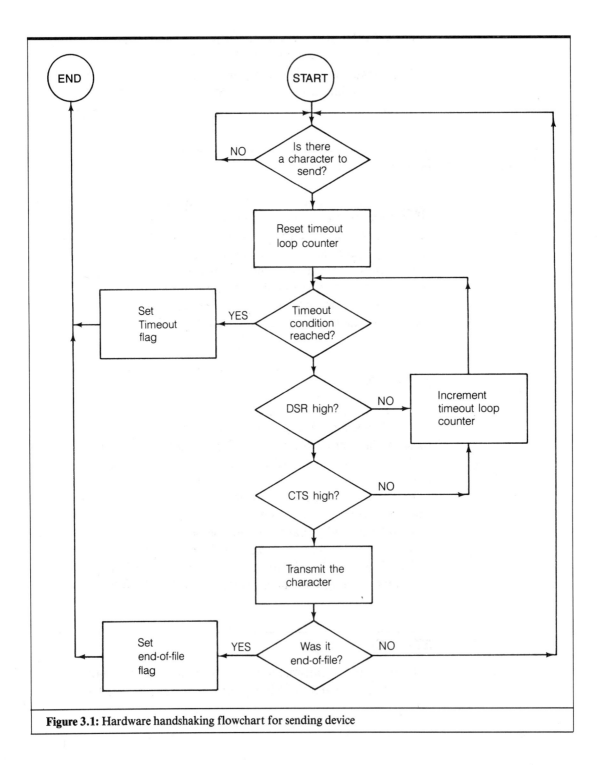

Figure 3.1: Hardware handshaking flowchart for sending device

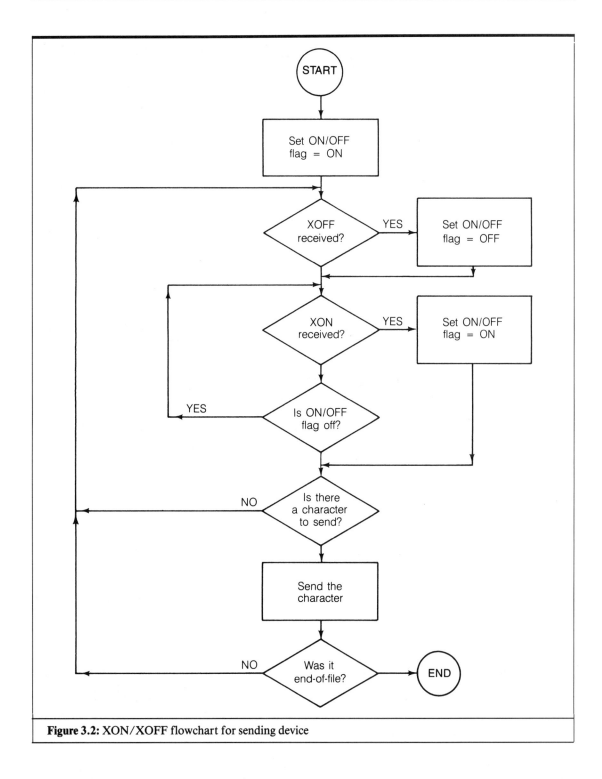

Figure 3.2: XON/XOFF flowchart for sending device

device sends an ETX (end of transmission) character, ASCII 3. The receiving device acknowledges receipt of the transmission by sending an ACK character, ASCII 6. Sometimes a NAK (negative acknowledge) character, ASCII 21, is sent back by the receiving device to indicate that errors were detected.

COMBINED HARDWARE AND SOFTWARE HANDSHAKING

*I*magine that you are using a personal computer or a terminal to communicate with a mainframe computer via a modem. Your modem probably uses hardware handshaking with your computer. The mainframe's modem might be using hardware handshaking with the mainframe. But the mainframe uses software handshaking with your computer.

Accordingly, your computer must be programmed to communicate only when the DSR line from the modem is high (and possibly only if carrier detect, or CD, is high) and a software stop signal has not been received. This can complicate the flowchart. However, some computers take care of the hardware handshaking automatically, and wait for the hardware handshaking signals to be high before sending a character, so that your program need only deal with software handshaking.

BUFFERS

A buffer is an area of memory into which received characters or characters to be transmitted are placed. The use of a buffer reduces the number of handshaking signals that must be sent because data can be transmitted in large blocks rather than character by character.

INPUT BUFFERS

An *input buffer* is used when the receiving device is receiving characters faster than it can deal with them. For example, a printer might be receiving characters at 1200 baud but only printing them at the equivalent of 300 baud. Rather than have the printer instruct the sending computer to stop after each character until it has been printed, the printer designer often sets aside an area of memory within the printer that has the capacity to store a given number of characters.

This area of memory is known as an *input buffer.* It is possible to think of this buffer as a water tank. The tank is being filled at the top, and at the same time it is being emptied at the bottom. A stop signal is sent when the buffer is almost full. The restart signal is sent when the buffer is almost empty. If the printer waited until the buffer was completely full before it said stop, and said start as soon as there was any room at all, the buffering would be defeated as soon as the buffer was filled for the first time. From then on it would be saying stop after each character was received and start after it had been processed, just as if there were no buffering at all.

Another reason for sending the stop signal before the buffer is completely full is to avoid losing characters that might be received simultaneously with the stop signal.

If hardware handshaking is in effect, a stop signal usually causes the sending device immediately to suspend transmission. With software handshaking, however, there is likely to be a time delay before the stop command takes effect, because the stop command has to be processed by the sending machine and characters can be sent out concurrently with this processing.

The next three figures illustrate various stages in the filling and emptying of a printer buffer. Figure 3.3 shows the buffer half full, at which stage it is both receiving characters from the computer and sending characters to the printing mechanism. Figure 3.4 shows the buffer almost full, at which stage the printer asks the computer to stop sending data by lowering the handshaking lines or sending XOFF. Figure 3.5 shows an almost empty buffer; at this stage the printer orders the computer to resume sending characters by raising the handshaking lines or sending XON.

OUTPUT BUFFERS

Output buffering refers to an area into which data are placed prior to being transmitted. This reduces operator inconvenience.

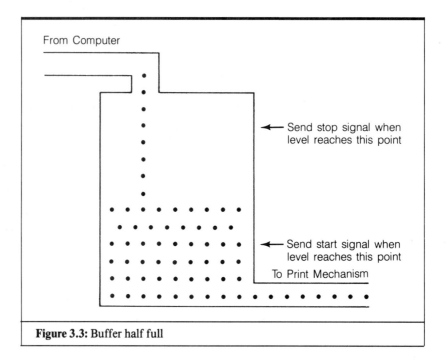

Figure 3.3: Buffer half full

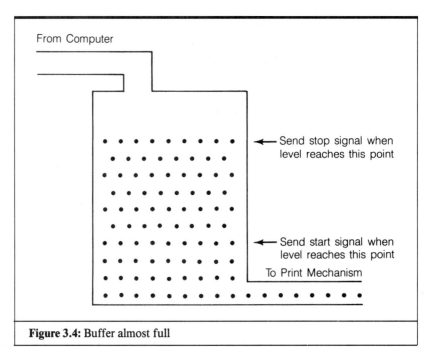

Figure 3.4: Buffer almost full

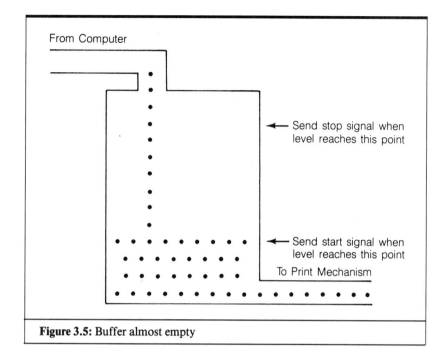

Figure 3.5: Buffer almost empty

For example, imagine you are typing at the keyboard, and the characters you type are being sent directly to a printer or other device. When the printer has received all the information it can handle and sends a signal to stop, you actually have to stop typing. With buffered output, however, you can continue typing until you fill the buffer, and by that time the printer will probably be ready to send a start signal again. In practice, most computers have a keyboard input buffer also, into which characters are placed as they are typed. Programs then take their input from the keyboard buffer.

IN-LINE BUFFERS

It is possible to purchase devices that stand between a computer and a printer and contain a large buffer (sometimes storing as much as 64,000 characters). They receive characters from a computer and send them to a printer. They receive data much more quickly than a printer normally does, and can send them on to a printer at the appropriate baud rate for the printer.

From the point of view of the computer, it is just sending data to a very fast printer. From the point of view of the operator, the operation is complete as soon as the data have been sent to the buffer, and he or she can continue creating a new document while the first one is still being printed out.

Some sophisticated in-line buffers can handle extra tasks such as conversion from serial to parallel, switching between printers, printing multiple copies of a document, and storing data received by a modem for subsequent processing by a computer.

One such buffer, the Hayes Transet 1000, can act as an answering machine for your computer. By connecting it to a modem you can have your modem answer the telephone and store the received data in the buffer even if your computer is switched off. It also formats, page numbers, and date stamps your printed material, enables two computers to share a printer or one computer to access two printers, and does a number of other tricks.

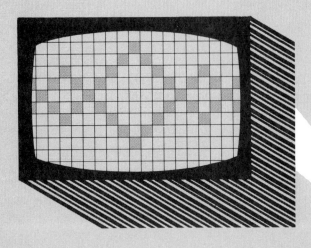

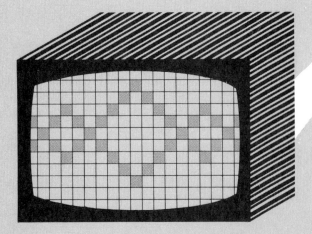

CHAPTER 4

Modems

WHAT MODEMS DO

A modem transforms data received from a computer in serial form into a form suitable for transmission through the telephone system, and vice versa. Modems can be used for transferring data between two computers at remote locations, or for connecting a terminal to a computer, in circumstances where a direct connection is not possible.

We have already discussed how data are passed between two devices connected directly through a serial connection. Interposing two modems and a telephone wire does not affect the baud rate, the number of data bits, parity bits, stop bits, or software handshaking. What the modem does is convert the plus and minus voltages that represent the individual bits of each character into signals appropriate for telephone communications. Different modems work in different ways; the most popular types of modems are described in this chapter.

HOW MODEMS WORK

C onnecting two computers, via modems, to the telephone system is by no means the whole story. As far as the user is concerned, the message disappears into the telephone system at one end and reappears out of it at the other end. Meanwhile, the signal may have passed through the cables of the originator's local telephone company to a long-distance telephone company, been bounced up to that company's satellite and back, and passed to the recipient via another local telephone company.

In the case of communication with another person on-line via CompuServe, all the above may take place, with the additional intervention of two network operators (and maybe their satellites) and CompuServe's mainframe computers.

Fortunately for us, we can use one of the established protocols for modem communication discussed below and, in most cases, can forget about the rest. Other protocols exist, but these are the ones in common use in the United States.

300 BPS MODEMS

When the plus and minus voltage signals are sent from the transmitting computer to a 300 bps modem, they are converted into tones. One tone is used to represent a positive voltage, and another to represent a negative voltage. In order for two-way communication to be possible, four different tones are used: two for the *originating* modem (the one that starts the communication) and two for the *answering* modem. The four tones used are defined in the Bell 103 standard, and are listed below:

- Originator transmits logic 0, positive voltage: 1070 Hz
- Originator transmits logic 1, negative voltage: 1270 Hz
- Answerer transmits logic 0, positive voltage: 2025 Hz
- Answerer transmits logic 1, negative voltage: 2225 Hz

This technique is known as *Frequency Shift Keying,* or *FSK.*

1200 BPS MODEMS

Modems communicating at 1200 bps generally follow the Bell 212A system. (Note that modems that allow operation at 300 and 1200 bps actually include two modems, one following the Bell 103 standard and one following the 212A standard.) Instead of using four discrete tones to represent the individual bits, they use *phase modulation* techniques. A carrier signal is phase modulated in various ways to represent different combinations of bits.

It is beyond the scope of this book to explain the electronic theory underlying these techniques. But the principle involved is that a carrier signal can be changed in certain ways so that two or more bits of information are transmitted at the same time. If a carrier signal can be in one of four different states at a given time, it can be giving two bits of information at once.

For example, if we call the four states A, B, C, and D, state A could represent 00, state B 01, state C 10, and state D 11. This effectively doubles the amount of information given for any frequency of change of state. Although technical factors limit the number of times per second the state can change to 600, with four possible states instead of two, 1200 bits of information per second can be transmitted.

The four states are achieved by the use of four phase angles, and the technique is known as *Phase Shift Keying,* or *PSK.* Two different carrier signals are used: the originator uses 1200 Hz and the answerer uses 2400 Hz.

2400 BPS MODEMS

In order to achieve a higher rate of information transmission, the number of states of the carrier signal can be increased again. Under the V.22 bis protocol, twelve phase angles and three amplitudes are used. This is known as *Phase Amplitude Modulation,* or *PAM.*

PAM yields 36 different states, which can produce six bits of information at each change of state, or 3600 bps at 600 changes of state per second. Certain limitations, however, are placed on which different states can be adjacent to each other; not only does this enable error-checking to be incorporated, but by the use of sophisticated algorithms an intelligent guess can be made by the receiving modem as to the contents of damaged data. As a result, the error levels at 2400 have been found to be similar to those at 1200 for similar telephone line conditions. A 2400 bps modem designed for an IBM PC is shown in Figure 4.1.

HIGHER SPEED MODEMS

By combining phase and amplitude modulation you can have rates of up to 9600 bps. However, these techniques are not yet widely used for microcomputers because no generally accepted standards have emerged.

There is a new type of modem that is becoming available called *FASTLINK.* It uses up to 512 different carrier tones simultaneously, and automatically adjusts its rate of transmission depending on the quality of the line. It even has built-in error checking so that messages that are received incorrectly can be automatically

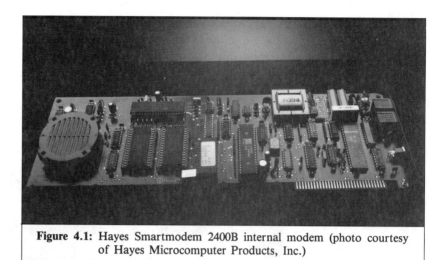

Figure 4.1: Hayes Smartmodem 2400B internal modem (photo courtesy of Hayes Microcomputer Products, Inc.)

retransmitted. Speeds of up to 15000 bps have been claimed over ordinary telephone lines. However, since the modem does not use generally accepted protocols at these higher speeds, FASTLINK modems must be in use at both ends of the communication for the high speeds to be achieved.

CONNECTING THE MODEM TO YOUR COMPUTER

*A*s far as your computer is concerned, an external modem is just another serial device. The normal wire connections are made for transmitted data, received data, ground, and handshaking. In addition, you can also connect Carrier Detect (CD), line 8, so the modem can let the computer know when a carrier signal is present, and Ring Indicator (RI), line 22, so the modem can indicate that the telephone is ringing. A positive voltage on CD means that a carrier signal is present. A positive voltage on RI means that the telephone is ringing. Whether these signals are actually recognized by the computer depends on the communications software in use. Some modems send messages on the data line when the carrier is lost, in addition to lowering CD, and send messages when the telephone is ringing as well as raising RI.

In the case of an internal modem such as the Hayes Smart-modem 1200B, the modem is contained in a card that is plugged directly into the computer. The card appears to the computer to be a serial interface card, so that software does not need to know whether an internal or an external modem is in use. Figure 4.2 shows the installation of an internal modem.

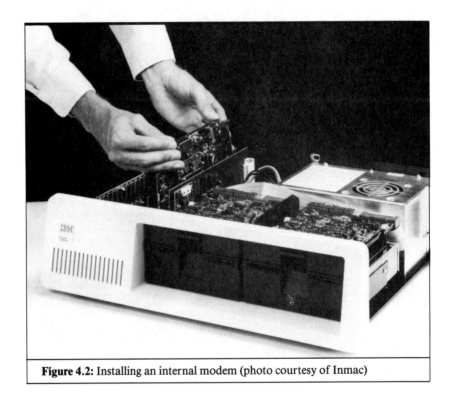

Figure 4.2: Installing an internal modem (photo courtesy of Inmac)

Note that hardware handshaking signals (DTR, DSR, etc.) are used to control communication between the computer and the modem. They are not passed along the line to the remote modem and computer. Accordingly, to enable handshaking between the two computers, software handshaking must be used.

CONTROLLING THE MODEM

*T*o use older types of modems, you have to dial the number using a conventional telephone, and then switch out the telephone and switch in the modem. Modern modems, on the other hand, usually do the dialing for you, and have many other built-in facilities that are discussed below.

Because the modems made by Hayes Microcomputer Products, Inc. have been so widely accepted by microcomputer users, the facilities and commands used by them have become a de facto standard in the industry. This has led competing modem manufacturers to follow suit, and many of them advertise their products as "Hayes compatible." Even IBM sells a Hayes compatible modem! The following information applies to the Hayes Smartmodem 1200 and 1200B modems, and should also apply to "Hayes compatible" modems if they are truly compatible.

COMMAND MODE AND ON-LINE MODE

The modem is always in one of two states: local command state or on-line state. While the modem is in local command state, instructions can be given to it from the computer (in other words, through the serial interface between the computer and the modem). These commands are, for example, to dial a number, or to answer automatically when the telephone rings. The commands are diverted to the modem, and not transmitted.

Once connection is established with a remote modem the local modem enters on-line state and no longer attempts to interpret the data being sent to it, but instead transmits it. If the carrier signal is lost, for example, because the remote modem has hung up the phone, the modem will revert to local command state. It is possible to return from on-line to command state without disconnecting by waiting for a *guard time* (the default is one second), typing an *escape command* (the default is +++) and waiting for one second before sending data to be interpreted as commands.

The instructions sent by the computer to the modem can be sent by communications software, or can be typed at the keyboard provided the keyboard output can be redirected to the serial port. Instructions sent in command mode should be sent with seven data bits and one parity bit, or eight data bits and no parity bit. There should be one stop bit unless you are communicating at 110 bps, in which case there should be two stop bits. Hayes-compatible modems can detect rate automatically.

RESULT CODES

When the modem receives a command it returns a result code. This code can either be in the form of a text message, or a numeric code. If you are controlling the modem through software, a numeric code is more appropriate. If you are controlling the modem from the keyboard, a text message is preferable. You can set the type of result code you want by using a command or by using switch settings. Table 4.1 shows the result codes.

COMMAND LINES

Command lines all start with AT or at (not At or aT) unless otherwise specified. The modem can detect the baud rate, word length, and parity from these two characters. You must already, of course, have set up your computer with the appropriate parameters. Several commands can be given in one command line.

Dialing Commands

A comprehensive set of commands that tell the modem to dial numbers is provided with the manual accompanying the modems. Table 4.2 lists a summary of these commands.

Following are some examples of how to use the dialing commands. Say you want to dial a number using Touch-Tone. Type the following:

ATDT1234567

Digit code	Word code	Meaning
0	OK	Command executed
1*	CONNECT	Connected at 0–300 bps
2	RING	Ringing signal detected
3	NO CARRIER	
4	ERROR	Error in command line
5*	CONNECT 1200	Connected at 1200 bps
6	NO DIALTONE	
7	BUSY	
8	NO ANSWER	
10	CONNECT 2400	Connected at 2400 bps

* With the 1200 bps modem, if X0 (the default) is set, result code 1 is given for both 0–300 bps and 1200 bps connections. This is to ensure compatibility with software written for 300 bps modems.

Table 4.1: Hayes Modem Result Codes

Command	Meaning
ATDT	Dial using touch tones
ATDP	Dial using pulse tones
,	Pause between numbers each side of a comma
ATT	Use touch tone as default
ATP	Use pulse as default
!	Transfer call
W	Wait for a second dial tone
@	Wait for one or more rings followed by five seconds of silence
O	(at end of command line) Return to on-line state
;	(at end of command line) Stay in command mode after executing command
R	Call an originate-only modem (see text)
/	Wait 1/8 second

Table 4.2: Hayes Modem Dialing Commands

To dial a number using pulse tones, type

.   ATDP1234567

After either of the above examples is typed, the modem dials the number and waits for a carrier signal. If it doesn't receive the carrier signal during a given time (the default is 30 seconds) it hangs up and returns a NO CARRIER result code.

If you need to dial 9 to get an outside line before you dial the number, type the following (the comma indicates a pause):

ATDT9,1234567

To dial 9 for an outside line using Touch-Tone, pause, and then dial the number using pulse dialing enter

ATDT9,P1234567

To dial 9 for an outside line, pause, dial a number, pause, and then transfer it to another number using #7, type

ATDT9,1234567,!#71234

To dial an MCI or Sprint number, wait for a second dial tone, and dial another number, type the following (the spaces do not do anything but are included for clarity)

ATDT 123 1234 W 123 123 1234

To call an originate only modem you must type R at the end of a dialing command line. You will recall that by convention the originating and answering modems use different frequencies. Typing R causes the modem to use answering frequencies even though it is originating the call.

ATDT 123 1234R

ATX Commands

Although the term "Hayes compatible" is often used, there is actually no absolute standard since not even all Hayes modems work the same way. The newest Hayes modems have facilities such as the ability to report connection at different baud rates, detect the busy signal, and detect the dial tone. Since older software does not recognize the result codes returned by these new facilities, the default mode of the newer modems is not to use them. The new codes have to be enabled specifically with the ATX commands. The idea is that newer software that can handle the newer codes will issue the ATX command indicating which codes it can deal with.

The 300 bps models send the string CONNECT, or numeric code 1. The 1200 and 2400 bps modems send CONNECT 1200 or CONNECT 2400 (numeric codes 5 or 10) as appropriate. Software that was written for the older modems might not recognize the new codes. Accordingly, if ATX0 is issued or if no ATX command is issued at all, the modem will return the same code as a 300 bps modem. If any of the ATX1 through ATX4 commands are issued, the modem will return a connect code appropriate to the bps rate. In addition, command ATX2 enables the NO DIALTONE result code, ATX3 enables the BUSY result code, and ATX4 enables both the NO DIALTONE and the BUSY result codes. The ATX commands are summarazed in Table 4.3.

ATXx	Code for 1200 bps	Wait for dial tone	Error if no dial tone	Error if busy
0	1	No	No	No
1	5	No	No	No
2	5	Yes	Yes	No
3	5	No	No	Yes
4	5	Yes	Yes	Yes

Table 4.3: Hayes Modem ATX Codes

Other Commands

There are quite a few remaining commands listed in Table 4.4. Most are self-explanatory. You will see that the ATS0 through ATS16 commands refer to setting modem registers. These registers record various parameters such as timing. You can set a value into a register by issuing the command ATS followed by the register number and the value. You can find out what the current contents of a register are by using the command ATS followed by the register number and a question mark. Table 4.5 shows the commonly used registers.

Command	Meaning
ATH	Hang up
ATZ	Hang up and reset to default settings
A/	Repeat last command
ATB0	Use international protocol
ATB1	Revert to bell mode
ATC0	Turn off carrier signal
ATC1	Turn on carrier signal
ATE0	Turn off echo to screen
ATE1	Turn on echo to screen
ATF0	Turn off half duplex
ATF1	Turn on half duplex
ATL1—3	Set speaker volume
ATM0	Turn off speaker.
ATM1	Turn on speaker until connected
ATM2	Turn on speaker and leave it on.
ATQ0	Turn on result codes (the default).
ATQ1	Turn off result codes.
ATV0	Display result codes as digits.
ATV1	Display result codes as words.
ATY1	Send four seconds of break signal before disconnecting. Disconnect if 1.6 seconds of break signal is received.

Table 4.4: Other Hayes Modem Commands

Command	Meaning
ATY0	Neither send nor respond to break signals (the default).
ATH1	Operate the telephone line relay and auxiliary relay.
ATH2	Operate the line relay.
ATS0—ATS16	Set the modem registers

Table 4.4 (cont.): Other Hayes Modem Commands

Register	Range/Units	Description
S0	0–255 rings	Ring to answer on
S6	2–255 seconds	Wait time for dial tone
S7	1–60 seconds	Wait time for carrier
S13	bit mapped	UART status register

Table 4.5: Hayes Modem Commonly Used Registers

Answer Mode

The modem can be set to answer the telephone automatically. You can regulate the amount of times the phone rings before the modem answers by setting register S0. For example, if you want the modem to answer on the fifth ring, type

ATS0 = 5

The default is S0 = 0, which tells the modem not to answer. There is a switch that can make the default S0 = 1, which means it will answer on the first ring unless instructed otherwise.

When the modem answers the telephone, it sends a carrier signal and waits for a response. If no carrier signal is received within the time set in register S7, it hangs up.

Detection of incoming baud rate is automatic, and the modem returns a result code indicating the baud rate. Your computer has to recognize this result code and adjust its own baud rate in order to respond to different incoming baud rates automatically.

PROGRAMMING THE MODEM

Commands sent to a modem can be entered at the keyboard or sent from a program. The modem, of course, cannot tell the difference. If you are using a program, you should probably turn off the text reporting of result codes, so that instead of reporting the results to you, the modem reports directly to the program which, in turn, interprets the codes itself.

Next, you should read the contents of all the registers and save them. This way, you will be able to restore the modem to the state it was in before the program started.

You can see that the powerful commands built into the modem make completely unattended operation possible. You can write programs that tell the modem to dial a particular number in the middle of the night, upload and download data, and so on. This allows businesses to take advantage of cheaper communication rates at off-peak hours, and a company can have all its branches dial in automatically with the day's figures every night, or exchange electronic mail, completely automatically.

By using an intelligent communications buffer, a lot of this can be done without even using a computer. The Hayes Transet 1000, when used in conjunction with a modem, and the Prometheus ProModem with its buffer options, can even store passwords, and are examples of how modems are becoming smarter all the time.

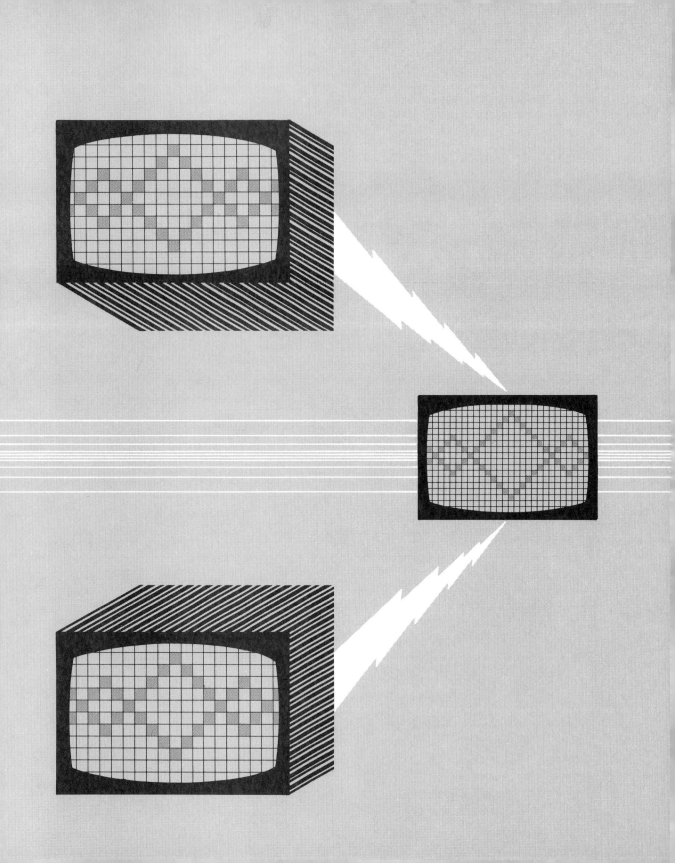

CHAPTER **5**

Telecommunications
Methods

INTRODUCTION

*M*ost of us think of telecommunications in terms of picking up a telephone and dialing a number. The computer equivalent of this, using an auto-dial modem on one end and an auto-answer modem on the other, is the simplest method. There are, however, many other alternatives, and I will review some of these methods in this chapter.

DIRECT-DIALED CIRCUITS

*T*he simplest way of communicating between long distance computers is to hook up a conventional modem and dial the number of the remote computer which will have been set up to answer the call automatically. This is known as direct dialing. In the United States, your call passes through the local telephone company, a long-distance carrier, and the local telephone company operating where the destination computer is located. You can choose your long-distance carrier from any of those available today, although the largest and most popular is still AT&T Communications. In most countries other than the United States, the call is handled by a single carrier—normally a public utility controlled by the government.

WATS (Wide Area Telephone Service) is a method of purchasing connect time in bulk. You pay a predetermined amount for unlimited use of a line for a given time period over a given distance. The arrangement can apply both to inbound calls, generally via toll free 800 numbers, and to outbound calls.

Despite their simplicity and directness, there are a number of problems with the direct-dialed circuits. First, the cost of the call is often higher than other methods. Second, the quality of the line is often insufficient for error-free communication. Third, high speed communications cannot be achieved over regular circuits.

LEASED LINES

*W*hen a large number of calls are made to one number, it can be worth installing a *leased line*. This is often done when, for example, a company's branch offices require constant access to a central computer. With leased lines, often called *point to point* connections, the customers don't need to dial. They simply pick up the phone and are automatically connected to the central computer. The customer generally has exclusive use of a particular circuit, although recently, *virtual leased lines* have been introduced in which one shares the physical circuits with other customers but any one customer still has the impression of exclusivity.

VOICE GRADE

The quality of a leased line connection depends on several factors. Most local telephone circuits, between the user and the local telephone company, consist of two wires. These wires carry communications in both directions. Four wires are necessary for longer distances because amplification is required, and a two wire circuit working in two directions at once cannot be amplified without first being converted into a four wire circuit.

As we saw in Chapter 4, modern modems are capable of operating in both directions at the same time on a two wire circuit because different frequencies are assigned to the orginator and the answerer. However, some modems are designed for use over leased lines or private circuits, and these modems can take advantage of a four wire connection by using each pair of wires in one direction only. In order for this to happen, a four wire connection with the local telephone company must be installed. Accordingly, you may come across both two and four wire leased lines.

Regular telephone circuits, as we all know, are susceptible to noisy lines. This noise disrupts data communications, because the modems sometimes interpret it as signals. It is possible, for an extra cost, to arrange for a higher than normal quality of line.

This service is known as *conditioning,* and is provided by the long distance carriers. AT&T, for example, offers several different levels of conditioning. With each higher level of conditioning the percentage of errors goes down and the cost goes up. Conditioning is achieved by using higher quality circuits and by the use of electronic methods of filtering out noise and increasing the strength of the signal.

DIGITAL CIRCUITS

If your communication requires higher speed or multiple circuits, a private digital circuit can be installed. These circuits transfer digital data synchronously at up to 56,000 bits per second. With digital circuits, modems are unnecessary, because the data are in digital form throughout the transmissions. Each circuit can carry a number of concurrent logical connections—in other words it can carry several transmissions at the same time.

A digital system can be installed in such a way as to appear transparent to the user or even the applications programmer. The terminals in an office can be connected to a *multiplexer,* which appears to the terminals just like the mainframe computer itself, although the mainframe could be located on the other side of the continent. The multiplexer then communicates, via a high speed digital line and a communications controller, with the mainframe computer. This technique is sometimes referred to as *multipoint* or *multidrop* and is illustrated in Figure 5.1.

PACKET SWITCHING NETWORKS

Value added network operators such as Telenet and Tymnet offer a service to people with long distance computer communications requirements that aren't extensive enough to justify the installation of expensive equipment and circuitry. These less expensive networks allow a user to access a remote database, for example, by dialing up the local *node,* or access point, of a value

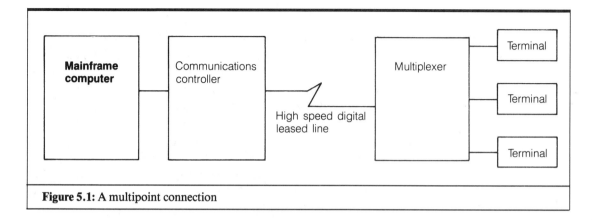

Figure 5.1: A multipoint connection

added network. The network then completes the connection to the remote computer. This is not only cheaper than making a long distance telephone call direct to the remote computer, but also results in a better quality connection because the network circuits are optimized for computer, rather than voice, communications.

A major difference between the services provided by a long distance carrier for telephone communications and value added networks is that the destination computer is charged for the service, rather than the originating computer. The charges are, of course, passed back to the subscriber when an information source is being accessed.

These networks are called *packet switching networks* because they use a method called packet switching to transmit messages. Packet switching involves dividing a message up into individual packets, and delivering the packets to an addressee that reassembles them back into the original message. Equipment for assembling and disassembling packets is known as a *PAD* (Packet Assembler/Disassembler).

The packet switching networks use high speed digital links, which can be land lines or satellite communications. They use synchronous communications, usually with the X.25 protocol. The routes are continually being optimized, and successive packets of the same message need not necessarily follow the same path.

Each packet consists of some sort of header, the body of the message, an error checking code, and an end of packet indicator.

Packets are not all of the same length, and can sometimes consist of just a single character if data are being received slowly. The PAD can be instructed to wait for a certain number of characters before transmitting a packet, or to wait for a fixed length of time. The method chosen is important, since sometimes a per-packet charge is made. If a block of data is to be sent, clearly it would be advantageous for it to be sent in a single packet rather than as sequence of packets each of one character.

Connection with the packet switching networks is normally made through regular dialed circuits, but direct connections can be established. The packets can be constructed by the user or by the local node. Sometimes a charge is made for packet construction. The X.PC protocol (see below) is a method by which packets can be constructed on a regular microcomputer, and sent to the node in asynchronous form. The node will then convert the packets into synchronous X.25 format.

Despite the complexity of packet switching, you do not need to know any details about how the system works if you want to communicate with an information provider or utility such as CompuServe or Dialog via a network such as Tymnet. Just dial the local node using a regular asynchronous modem. You will then be asked to enter a terminal identifier code and a code for the host computer you wish to access. You may think that you are directly connected to the remote computer, but the echoing of characters is normally done by the local network. Otherwise, the communication would take much longer because the time delay for characters to be sent to the remote computer and back can be considerable, especially if the network is using satellite links. Figure 5.2 illustrates the use of Tymnet to access CompuServe.

Packet switching is internally more complicated than direct dialing, but due to its lower costs and better quality connections, it is very popular. The fact that over half a million people regularly use the networks to access information providers such as CompuServe, the Source, Dow Jones, and Dialog is testimony to their effectiveness and ease of use.

The following is a list of some major packet switching networks:

- ARPANET, the U.S. government network
- Net 1000, the AT&T network

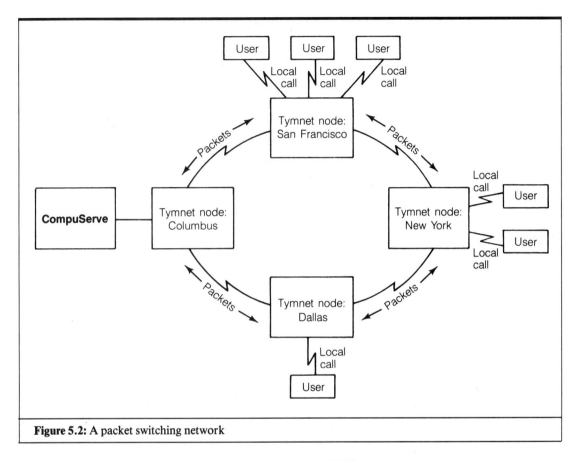

Figure 5.2: A packet switching network

- Telenet, operated by GTE
- Tymnet, operated by McDonnell-Douglas
- Safelink, operated by Western Union
- Uninet, operated by United Telecommunications and Control Data Corporation

THE X.PC PROTOCOL

*T*ymnet (pronounced Timenet) developed the X.PC protocol. It is new and not yet widely adopted, but it offers considerable advantages in facilitating the connection of asynchronous devices to the packet switching networks. It is an asynchronous protocol, based on a subset of the X.25 protocol, which was designed for synchronous communications. Although designed for network communications, X.PC can also be used as a file transfer protocol in direct connection between computers.

The main features of X.PC are described below. For more information you can contact the following address:

> Tymnet, McDonnell Douglas Network Systems Co.
> 2710 Orchard Parkway
> San Jose, CA 95134

LOGICAL CHANNELS

X.PC not only provides a comprehensive error-checking protocol with automatic retransmission, it also allows several sessions to take place concurrently through one modem and one telephone line. Through Tymnet, many different host computers can be accessed from a single terminal or microcomputer and there can be up to fifteen *logical channels,* each of which can be one-way incoming, one-way outgoing, or two-way.

Normally, when logging onto a remote host via Tymnet, you dial up the local Tymnet number and enter a code for the host computer you are accessing (for example, CompuServe). Using X.PC, you could be connected to several hosts at the same time, and if you are using a multitasking or windowing program, you could have several different sessions going on at the same time.

For example, an executive could be watching up-to-the-minute NYSE quotes running across the top of his screen, while pulling a file from the corporate database, researching Dow Jones' QUOTES, comparing securities portfolios recommended by Bridge Data's BROKERAGE SYSTEM, and, at the same time, his E-MAIL messages could be printing.

Alternatively, several computers connected through a local area network could share a communications server and a modem and all be connected to different host computers through one telephone connection to Tymnet.

ISDN

A standard known as ISDN (Integrated Services Digital Network) is becoming adopted that will simplify digital communications of various types (data, voice, video, and others). Pacific Bell at the time of writing (February, 1986) is testing a system known as *Project Victoria*. With this system, five data and two voice channels can be transmitted simultaneously over standard telephone twisted-pair wiring or fiber optic cables. Any standard transmission speed from 9600 bits per second to 300 bits per second can be selected for one data channel, while four other data channels will each accommodate speeds up to 1200 bits per second. All that a user needs is a decoding unit attached to the regular telephone connection socket. Telephones, computers, and other devices are attached to the decoding unit.

It is envisaged that the spare data channels will be used to cover such services as emergency alarms and pay-per-view cable TV (not the TV channel itself, but the instructions requesting and enabling reception). An even more advanced possibility is the introduction of remote diagnostic facilities for domestic equipment whereby a computer at a central station could interrogate your television set or other electronic equipment and find out what is wrong with it. Telecommuting, or working from home and corresponding electronically, will also benefit from this development. The idea of each household being equipped with five serial interfaces offers exciting prospects.

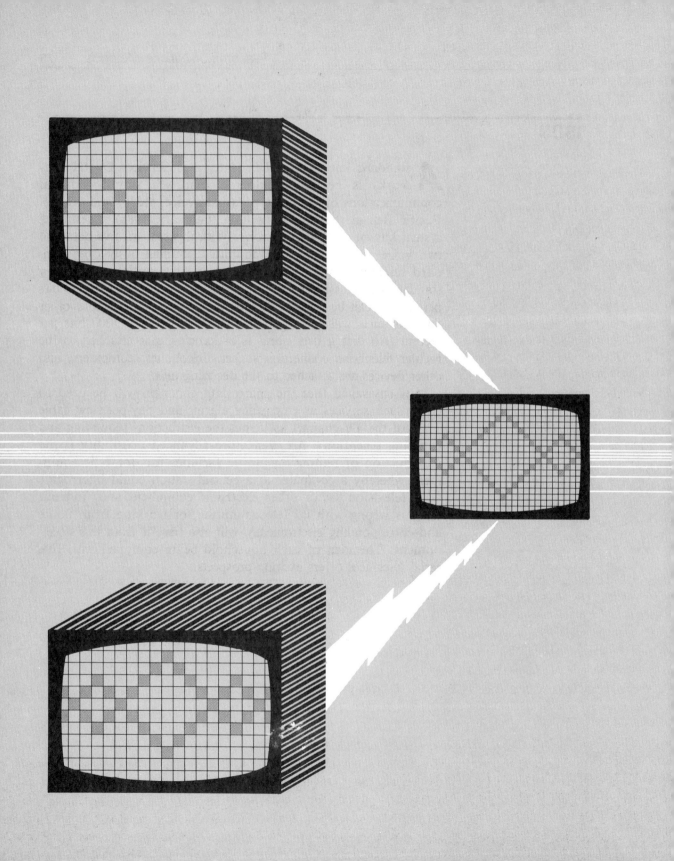

CHAPTER **6**

Micro/Mainframe Communications

THE GREAT DIVIDE

*T*ime and time again, when new inventions appear they are rejected by the establishment, and taken up by a band of individualists who are originally regarded as freaks. This happened with the automobile, with the stereo, and again with microcomputers. Once a technology becomes well established, the original freaks become the establishment and then mistrust and sometimes try to suppress the next new technology that appears.

It is particularly ironic in the case of the computer industry, which in theory ought to attract people who are interested in new technology. Yet, first the minicomputer industry and then the microcomputer industry were developed by new companies, not the established ones.

However, there is a difference. The minicomputer companies were started by people who left the mainframe companies. Microcomputers were mostly developed by people without large computer experience. This is one reason why there is still no really good operating system for microcomputers. All the same mistakes were made, and people did not benefit from the experiences of the past.

Because of the history of the microcomputer business there is still a great divide between the micro and mainframe worlds. As a result, it is hard to find people who have experience with both microcomputers and larger computers. It is easy to find people who can tell you the function of every circuit on the motherboard of an IBM PC, but who have never heard of SDLC. There are just as many DP managers who don't know how to copy a floppy disk.

Of course, the large companies have now moved into the micro world and largely taken it over—at least on the hardware side— and the two worlds are slowly converging. Executives who have microcomputers on their desks want to use them to access the corporate mainframe. I have seen, in a company that was changing mainframes, desks with a PC and two terminals on them. You can imagine how much room there was for anything else. There is enormous pressure to integrate micros into a unified DP system.

This chapter presents an introduction to the mini and mainframe communications world, and an overview of the ways in which microcomputers can participate in that world.

TERMINAL EMULATION

*M*ost communication between humans and mini and mainframe computers is with terminals. Terminals vary widely in their characteristics, but they are all primarily intended to take data entered at a keyboard or other human input device and transmit it to the host computer. There are several different types of terminals; a few of them are discussed below.

DUMB TERMINALS

Dumb terminals are terminals that depend on the host computer for all operations and do not do any processing of data on their own.

Teletype Terminals

The most basic type of terminal is the Teletype, which consists of a keyboard, printer, and punched tape device. Punched tapes can be created in local mode, edited if you are clever enough to know which holes to change, and fed back in to be read by the machine and transmitted to the host computer. Next time you complain about how long your PC takes to boot up, spare a thought for those of us who waited twenty minutes for the language to be read in from punched tape before we could start programming. We then communicated with the computer at the terrifying speed of 110 baud, with no screen and only the print-out to work from. Of course, it all seemed like magic at the time.

Video Terminals in Teletype Mode

The next most basic device is a terminal that has a screen, but no full screen processing. In other words, the text is displayed one line at a time, with the existing text scrolling up when the bottom of the screen is reached. This is known as *Teletype mode,* and basically treats the screen as a printer. PC-DOS works in this way, as do most information services. This enables them to be accessible from as many types of terminals as possible.

Full Screen Processing

The most advanced of the dumb terminals use *full screen processing*. With full screen processing, data can be presented at different places on the screen, rather than as a sequence of lines. The terminal has sufficient memory to store all the characters currently displayed (typically 24 or 25 lines of 80 columns) and their *attributes* (inverse, underline, blinking, intensity, and character set). By using special sequences of characters you can clear the screen, move the cursor to a position on the screen, select an attribute, and so on. These sequences are known as *escape sequences* because they generally start with the escape character (27 decimal, 1BH).

Unfortunately, different terminals use different escape sequences to achieve similar results. This means that software that is to be used on different terminals has to be configurable with the characteristics of the terminal in use at any one time. Most dumb terminals, however, conform to one of a small group of standards such as DEC VT100, Lear-Siegler ADM3A, and IBM 3101. In other words, they are either standard models produced by a major manufacturer or emulations of one of the major models.

INTELLIGENT TERMINALS

Intelligent terminals are so called because they are capable of processing information on their own. With the advent of cheap microprocessors it became logical to incorporate them into terminals. Many of the standard data entry tasks require very little computing power and it saves time and resources to handle these simple tasks within the terminal rather than use the resources of the mainframe computer.

Intelligent terminals offer various different features. One is called *block mode*. In block mode, the mainframe sends information to the terminal specifying fields to be completed. The terminal handles acceptance of the entry of data into the fields, and sometimes also carries out a certain amount of validation. Nothing is sent to the host computer until the user presses the Enter key. At this point all the fields are sent together in a block.

Forms caching is another feature of some intelligent terminals whereby the host computer can request the terminal to save one or more screens in the terminal's memory so that the saved screens can be redisplayed without having to be retransmitted to the terminal. Some terminals also offer a multiple session capability whereby the user can be logged onto different programs at the same time, and either switch between programs or view them simultaneously in different regions of the screen or *windows*.

The use of an intelligent terminal can speed up operator input by reducing waiting time, making better use of host computer resources, and making communications more efficient by reducing the number of individual transmissions.

MICROCOMPUTERS AS TERMINALS

Many terminals can be emulated using microcomputers and the appropriate hardware and software. To understand the advantages of using microcomputers as terminals, you only have to wander around the offices of a company that has a mainframe computer, and see the number of six foot by two foot desks that have both a microcomputer and a terminal on them. Because terminal emulation allows you to get rid of the terminal and work only with a microcomputer, it pays for itself in real estate alone.

A further advantage of terminal emulation is that you can download data onto a microcomputer, and analyze or otherwise process it locally. For example, you can retrieve sales figures from a corporate database and turn them into a pie chart using Lotus 1-2-3.

Unfortunately, downloading is not always easy. Data formats are not the same on all computers or with different software. Efforts are being made to solve this problem in two ways. Some software developers are producing mainframe and micro versions of the same product with special facilities for interchange of information. Also, attempts are being made to produce standardized data formats such as IBM's Document Content Architecture (DCA).

TERMINAL EMULATION PROBLEMS

The lack of data security is a major problem related to the downloading of corporate data. This is a constant problem for DP departments, and is often very little appreciated by the users. Whatever passwords and access restrictions have been incorporated into the log-on procedures to protect access to sensitive data, if the information can be saved onto disk it can potentially be stolen by, or inadvertently passed on to, unauthorized persons. This is one reason why DP departments often resist the use of microcomputers as terminal emulators.

Other specific areas affecting the feasibility of terminal emulation on a microcomputer are listed below.

First, the keyboard will almost certainly be different. There may be more or fewer function keys; there is probably only one key for both Enter and Return; other keys may be missing and may have to be simulated with awkward combinations of keys pressed together or in sequence. Second, the microcomputer may be slower, since it lacks certain dedicated circuitry (for example, scrolling) designed to optimize the performance of a terminal. Third, with color terminals the number and combinations of colors available may be different. Fourth, with graphics terminals, the number and scaling of the individual dots may be different and the dedicated graphics circuitry may be absent. Some effects can only be achieved with special chips and architecture. Fifth, the number of rows and columns may be different, entailing some sort of scrolling or other mechanism that makes the appearance of the screen different from that of the original.

Finally, there may be undocumented features of the original terminal that cause unexpected differences in performance; only trial and error can establish the exact way in which the real terminal behaves.

BEYOND TERMINAL EMULATION

The next step beyond fooling the mini or mainframe into thinking it is talking to one of its regular terminals is for the microcomputer to join the big boys' club, and participate as a computer rather than a terminal.

There are network architectures available that handle networks of larger computers such as System 36s and System 34s. One of these is *SNA,* or Systems Network Architecture. In some host environments, the APPC protocol (Advanced Program to Program Communications) enables microcomputers to join the SNA network and participate in parallel and distributed processing applications as well as communicate with each other using the advanced communications facilities of the network.

PHYSICAL CONNECTIONS

M any terminals communicate with their hosts using the RS-232 connections and modems we have already described in this book. When a microcomputer is emulating this type of terminal it merely has to incorporate a simple asynchronous adapter or internal modem as appropriate.

Terminals used with IBM mainframes, however, do not operate in the same way. Normally they are connected via coaxial cables to a cluster controller that is connected to the mainframe itself or its front end. The coaxial cable enables much higher data transfer rates to be achieved. In addition, these terminals use synchronous rather than asynchronous methods.

Many adapter boards are available for the IBM PC that enable it to emulate this kind of terminal. The boards have the appropriate coaxial connections for connecting with a cluster controller, synchronous circuitry, and appropriate software. Figure 6.1 illustrates a typical configuration using a PC in pure terminal emulation mode.

Other boards enable the IBM PC to emulate not only a terminal but also a cluster controller. These boards can connect terminals to an IBM mainframe via a modem and a communications

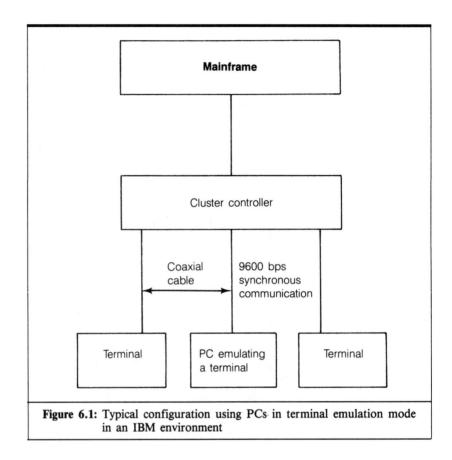

Figure 6.1: Typical configuration using PCs in terminal emulation mode in an IBM environment

controller. This method is useful if, for example, you need only one terminal and there is no cluster controller available. Data transfer rates using this method are generally limited to 2400 bps, which is slower than the rates at which a real cluster controller communicates; however, it eliminates the overhead caused by several terminals sharing a cluster controller. Figure 6.2 illustrates this alternative.

Finally, where personal computers are connected together in a local area network it is possible for a device on the network to act as a cluster controller and enable any other device on the network to communicate with the mainframe. This is illustrated in Figure 6.3.

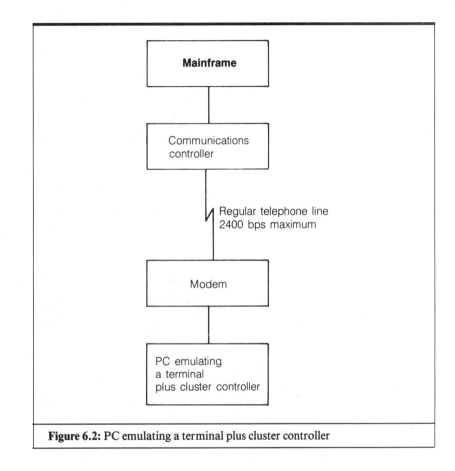

Figure 6.2: PC emulating a terminal plus cluster controller

SYNCHRONOUS PROTOCOLS

You will recall that with asynchronous communications, the most common format with microcomputers, individual characters are framed by start and stop bits and an optional parity bit. With synchronous communications, however, characters are sent in a continuous stream to the receiving device.

Various protocols have been devised for synchronous communications, requiring the transmission to be split up into messages, each with a start sequence and an end sequence. Some of the major protocols are discussed below.

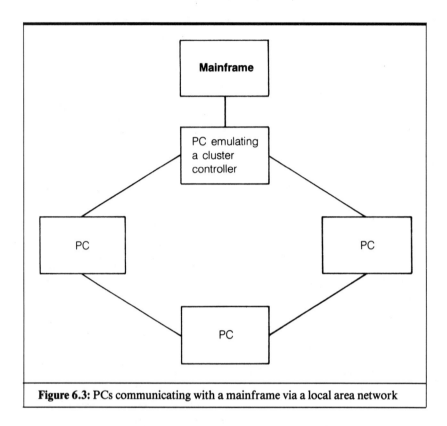

Figure 6.3: PCs communicating with a mainframe via a local area network

BISYNC

BISYNC (Binary Synchronous Communications Protocol), is a standard byte-oriented protocol used in IBM communications. As with asynchronous communications, BISYNC messages consist of a sequence of characters. Although IBM communications are almost always encoded in EBCDIC, ASCII and Six Bit Transcode are also available. These coded messages are divided up into frames, each of which consists of two sync bytes, an optional header (preceded by a start of header character and followed by a start of text character), the text of the message, a byte to mark the end of the text, and an error check number.

The byte marking the end of the text can be one of several different codes marking the end of the transmission block (ETB), an intermediate end of transmission (ITB) meaning there are more to come but here's an error check code to make sure we are all right so far, or an end of text (ETX) meaning we have now sent everything.

There are also defined control characters governing handshaking, marking the start of special control sequences, and indicating data that may resemble control characters but that are to be treated as part of the data stream.

SDLC

SDLC (Synchronous Data Link Control) is an IBM bit-oriented synchronous protocol. Messages are divided into frames, just as with BISYNC. However, there is no concept of individual characters or word length, as the communications are seen merely as a stream of bits. There is not even a fixed standard frame size. The receiving device merely watches for a special sequence of bits that signify the end of a frame.

Each frame consists of an eight bit beginning flag, an eight bit address field, an eight bit control field, a data field containing the information to be sent (any number of bits), a sixteen bit frame check, and an eight bit ending flag. The beginning and ending flags each consist of the binary sequence 01111110.

Since it is quite possible for the sequence 01111110 to appear by chance within a message, it is specified that after each sequence of five continuous binary ones within the message content, the sending device has to add a zero. The receiving device removes this zero. This avoids a sequence of six binary ones ever being sent except as part of a beginning or ending flag.

The control field contains information such as frame number, acknowledgement of messages received, and a message indicating whether the current frame is the last in the sequence. The frame check contains a 16-bit CRC computation as an error-checking device. CRC computations are discussed in Chapter 10.

Frames can follow each other with no gap in between. In this case the ending flag of the each frame doubles as the beginning flag of the next frame. Alternatively an idle sequence can be sent between frames to maintain synchronization. A sequence of eight consecutive ones serves as an abort.

HDLC

HDLC (High Level Data Link Control) is a protocol proposed by the International Standards Organization. It is based on, and similar to, SDLC.

ADCCP

ADCCP (Advanced Data Communications Control Procedures) is almost identical to HDLC. It is maintained by ANSI (American National Standards Institute).

X.25

X.25 is the standard protocol in use in packet switching networks, which were described in the last chapter. The frames transmitted under X.25 are similar to those in SDLC.

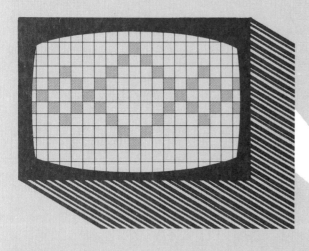

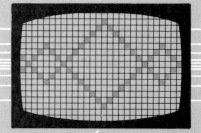

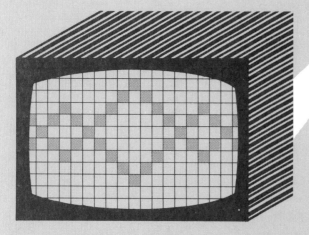

CHAPTER 7

File Transfer

INTRODUCTION

*I*n the course of working with computers you will often find it necessary to transfer files from one computer to another. Sometimes the receiving computer is intended to process the data in some way. For example, you may have written a letter on a portable computer and want to print it out with another computer that cannot read the same disks. At other times, the second computer is needed merely as a storage device for data that will ultimately be transferred onto a third computer. The use of a mainframe as a storage and exchange device is becoming more and more common. When you want to transfer a program or data to another user or to many users, you can upload the data to a central mainframe from which the other users can download it. This happens, for example, with CompuServe, when members of a SIG (special interest group) share their programs with other members. It also happens in universities as a means of distributing or transferring software.

Transferring files between computers can serve many purposes but can also pose many problems. This chapter focuses on the difficulties that can arise during file transfer. Several methods of solving file transfer problems have been devised. The following two chapters go on to discuss two of these: the XMODEM protocol, which was originally devised to facilitate transfers of data between microcomputers, and Kermit, devised to facilitate transfer between microcomputers and larger computers.

WHY PROTOCOLS ARE NEEDED

*S*etting up a computer to *transmit* data serially is rather straightforward; almost all computers offer the option of driving a serial printer, and can be programmed to send data out of a serial port without too much difficulty. Most larger computers *receive* serial data all the time because they are controlled through terminals that operate serially.

So you would think that in order to transmit data from a microcomputer to a larger computer one must simply make the microcomputer think that it is printing and make the larger computer think that it is connected to a terminal. However, it is not quite as simple as that.

The creators of the established methods of computer communication assumed that material would be transferred from a keyboard, in the form of text, and input at human typing speed. File transfers, on the other hand, often take place much faster than a person can type, and sometimes include characters that do not appear on the keyboard. The resulting problems are discussed below, most of which can be solved by the protocols discussed in Chapters 8 and 9.

WORD LENGTH

Much of the data contained in microcomputer files does not consist of text but contains computer programs, graphics data, or other non-ASCII material. These data, often referred to as *binary data,* normally use the full eight bits of each byte.

You will recall that the official ASCII table, which covers the most common keyboard characters, uses only seven-bit words. Many computer systems and communications channels, designed for ASCII input only, are unable to accept eight-bit words. They insist on seven-bit words, the eighth bit being used as a parity bit. Accordingly, binary data must be converted in some way to seven-bit words before many computer systems will accept them.

CONTROL CHARACTERS

Another problem you may encounter during file transfer involves control characters. Transferred data often contain special byte values that the target computer might misinterpret as ASCII or EBCDIC control signals. Some computers, in fact, will handle special characters in particular ways that cannot be overridden through applications software. In order to overcome this problem, the data must be converted into characters that can be accepted by the receiving computer. Kermit is an example of a protocol that can do this.

BLOCK LENGTH

With many computer systems, input is placed into a buffer before being processed. The size of this buffer is often limited, typically to 128 or 256 bytes. When this is the case, it is not possible to send a lengthy file as a continuous burst of bytes. The sending computer must divide up the file and send it as a series of blocks, each of which is smaller than the input buffer of the target machine.

Dividing data up into blocks also enables more sophisticated error checking to be carried out, as you will see later in this chapter.

HANDSHAKING

Another item to be aware of is whether different software handshaking procedures may exist on the two machines. For example, XON/XOFF might exist on one but not the other. The use of a file transfer protocol will help to overcome this problem.

ERROR CHECKING

We all experience noisy telephone lines from time to time. However, what appears as noise to us may appear as data to a computer. The longer the transmission and the higher the baud rate, the greater the likelihood of data being corrupted. Therefore it is always advantageous, when transferring files, to incorporate some form of error checking into your communication.

When the data consist of text, it is generally easy to see where errors have occurred because the text will look garbled. Where binary data are concerned, however, you can't usually tell what happened and you know only, for example, that your program does not run.

As we have seen in Chapter 2, parity checking is often included in serial communications. However, this check has only a fifty-fifty chance of detecting an error in a single byte, and a single error can be fatal to a computer program in machine-readable form.

Furthermore, it is annoying to be told that there has been an error, after transferring a long file, and have to transfer the whole file again when only part of it is corrupt. It is much better to divide the file up into blocks, and have each block checked for errors. That way, your computer will only have to retransmit the corrupted blocks and your telephone bills will be much lower.

Accordingly, most file transfer portocols divide the data into blocks, and check each block for errors.

FILE TRANSFER PROTOCOLS

*T*here are many different established methods for dealing with the above problems. Some of them are discussed below. Two of them, XMODEM and Kermit, are covered more fully in Chapters 8 and 9. However, if you are using a commercial communications package you may well find that a suitable protocol is included. Chapters 8 and 9 are necessary reading only for those of you who want to learn to write your own communications software.

HEX CONVERSIONS

A simple way to turn binary data into ASCII transmittable characters is to turn every byte into its hexadecimal equivalent and to transmit the ASCII characters corresponding to each number (i.e., the whole message will consist of the characters "0" through "9" and "A" through "F"): two characters for each byte. The message is reconverted by the ultimate destination computer (not necessarily the one being used to store the data).

This method solves the problem of sending eight data bits through a seven-bit channel, and the problems arising when control characters form part of the data. However, it does not solve the buffering, handshaking, and error checking problems mentioned above, and it doubles the physical length of the file and therefore the transmission time.

XMODEM

XMODEM is described more fully in Chapter 8. It was originally designed for transfers between microcomputers, but has been used in micro to mainframe transfers as well. XMODEM offers error checking, and the division of data into blocks. However, it does not offer the capability of sending eight-bit data through a seven-bit channel, or the conversion of control characters into printable characters.

KERMIT

Kermit is described more fully in Chapter 9. It is used primarily for transfers between mainframes and microcomputers and is widely accepted, especially in the academic world. It solves all the problems described in this chapter.

COMPUSERVE A AND B

These protocols, devised by CompuServe, are used in uploading and downloading data to and from CompuServe. They are offered in some commercial communications packages, as well as in CompuServe's own package.

X.PC

We mentioned the X.PC protocol in Chapter 5. Although not yet widely accepted, it is beginning to be incorporated into commercial packages. It is the most sophisticated protocol of the ones mentioned, and offers the advantage of close integration with X.25, which is the protocol used by most packet-switching networks. X.PC is really a communications protocol, but because it offers both error checking and division into blocks, it can be used as a file transfer protocol also.

OTHER FILE TRANSFER CONSIDERATIONS

*F*ollowing are a few additional items to keep in mind when you are transferring files.

PACKETS AND LAYERS

Both XMODEM and Kermit divide up files to be transferred into blocks or packets. Each packet consists of a header of some sort, the data themselves, an error check code, and an end of packet mark. The target computer is expected to send a response indicating whether the packet was correctly received. In the case of XMODEM, this response consists of a single ASCII character returned to the sending device. In the case of Kermit, the response itself is also in the form of a packet.

In Chapter 5, we mentioned the general use of packets in telecommunications. There is a big difference, however, between file transfer packets and telecommunications packets. When running a file transfer protocol such as XMODEM or Kermit, the target computer is aware of the packets, and a program running on the target computer analyzes each packet, checks the error code, strips out the data, and saves them to a file. When telecommunications systems use packets, the packets are generally transparent to each computer. The computers think that they are directly connected, one being a terminal to the other.

As a result, the packets being sent under a Kermit file transfer can themselves become the data part of packets being sent through a communications network. These packets can, in turn, become the data part of even higher level packets. The concept of structures within structures within structures is known as *layering*.

DATA FORMAT

The fact that it may be possible to transfer binary data from one computer to another does not necessarily mean that the second computer will be able to read the data. For example, the data may consist of programs that will not run on the target computer. Even if the two computers use the same processor chip, they may still be designed differently. And if the two computers are from the same family, the operating systems and memory requirements could still be different. Also, the data may have been created by a program that is not available on the target machine, or by one that is available but works differently on the two machines.

These complications don't matter when you are using the target machine as a storage medium. However, if you want the target machine to process the data in some way, you may have to convert the data.

Some programs, in particular some word processing, spread sheet and database programs, have the ability to read and convert data created by another program. For example, dBASE II and III and Lotus 1-2-3 are so prevalent that many software producers make their data formats compatible with those programs.

Attempts are also being made to establish standards for data storage formats, but there are still many cases where the data cannot be used even if they can be transferred.

HALF AND FULL DUPLEX

Half duplex means that both computers do not attempt to communicate at the same time. In other words, at any one time, one computer is in transmitting mode and one is in receiving mode. *Full duplex* means that the computers can both be transmitting at the same time.

When selecting a file transfer method one of the criteria should be whether or not both computers support full duplex. If they do support full duplex, the receiving computer can send a message concerning successful receipt of a block while receiving a later block. If only half duplex is available, the sending computer has to wait for an acknowledgment after sending each block. This procedure can easily double the transmission time, because waiting for a response to a block often takes longer than receiving the block.

Another difference between full and half duplex is that with half duplex the characters are not echoed back when they are received. This usually doesn't matter with binary data, but it can be useful to have some real-time evidence that the transmission is being received.

DISK FORMAT CONVERSION

It is sometimes necessary to transfer data between two machines that use the same size disks but format them in a different way. This is often the case with CP/M transfers. Sometimes this is treated as a serial communications problem, whereas, in fact, it can be solved with a simple format conversion program. Programs such as Xeno-copy, from Vertex, can read a wide variety of disk formats and convert them to other formats. Clearly this is not possible if the disks are of different sizes or types and the target machine cannot physically read them. But when both machines use five-inch floppy disks it is often possible to use one of these conversion programs.

Some companies offer a format conversion service that is often used when a company is converting from dedicated word processing machines to personal computers and has a large amount of data to transfer.

I mention the above alternatives just so that you will be aware that there are alternative solutions to file transfer problems and that transferring your data serially is not always the best method.

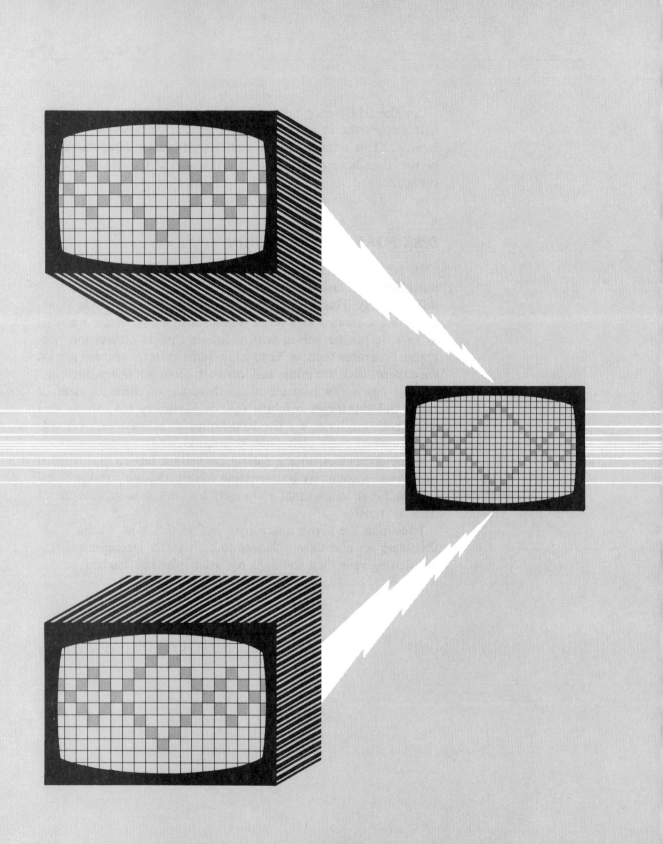

CHAPTER **8**

XMODEM

INTRODUCTION

I am indebted to Ward Christensen, the author of XMODEM, for the following short history, and for clarifying various aspects of the XMODEM protocol.

In September, 1977, Ward wrote a program for CP/M-80, called MODEM. He had incompatible disks so he wrote a program that allowed him to exchange programs with other CP/M users. The original program was intended to be used by two people, one at each computer, at opposite ends of the phone line. When David Jaffe invented BYE, which allows an unattended CP/M computer to be operated by another machine linked by modem, the need arose for an unattended version of MODEM. At this time, many people were revising MODEM itself, and it was taking on new version names such as MODEM7. While Ward's original MODEM program could be used on an unattended machine, the user had to remember to say Q (for Quiet mode). Keith Petersen stripped down MODEM and made XMODEM, which allowed transfer to and from unattended machines.

The protocol implicit in Ward's original program, which some called the "Christensen protocol," has become widely known as the XMODEM protocol.

XMODEM is used most often to download all types of binary and ASCII files from dial-in home computer systems, such as PCs, CP/M machines, etc. However, in spite of its heritage of micro-to-micro communications, XMODEM is also used in mainframe communications where the mainframe is capable of supporting it. CompuServe is a notable example of a mainframe communication system that offers XMODEM file transfers.

BLOCKS

D ata transferred by XMODEM are divided up into blocks. Each block consists of a start-of-header character (01H), a one byte block number, the one's complement of the block number, 128 bytes of data, and a one byte checksum. This format is illustrated in Table 8.1.

Offset	Contents
0	SOH (start-of-header character: ASCII 01)
1	Block number: starting with 1, but wrapping to 0 after FF
2	One's complement of the block number (255 − block number)
3 to	
130	128 bytes of data
131	Checksum: sum of the data bytes only, carry ignored

Table 8.1: XMODEM block format

The block number starts at one, but is computed modulo 256, meaning that after 255 (FFH) it goes back to zero. The one's complement can be computed by subtracting the block number from 255 or by complementing all the bits in the number (turning ones to zeros and zeros to ones). The checksum is a single byte computed by adding together the 128 data bytes and ignoring any carry.

FILE LEVEL PROTOCOL

*B*efore the sending computer can send data it has to receive a NAK (negative acknowledgment) character from the receiver. The receiver program is supposed to send a NAK character (15H) as a *timeout* after every ten seconds that go by (officially, but sometimes more often in practice) without receiving data. It is the first such NAK that triggers the transmitter to start sending.

Once the receiver program starts receiving a block, it reports an error whenever a gap of one second or more occurs between characters in the block, including the checksum. However, it must wait for the line to clear before sending a NAK to indicate an error.

Note that the one second timeout is not sufficient for many long-distance connections, and a longer wait is often substituted for the one officially specified.

The receiver then checks the block number and reports an error if it is out of sequence. If the block number is the same as the last one, it indicates a retransmission that should not be considered an error. After receipt of each block, the receiver sends ACK (06H) if the block is received correctly, or NAK if it is not. In the latter case, the transmitter resends the block. After the block is acknowledged, the *next* block is transmitted.

At the end of transmission, the transmitter sends EOT (04H) and waits for an ACK, resending the EOT if it does not get one.

THE CRC OPTION

*T*he one-byte checksum is not sufficient to detect all errors. Accordingly, an extension to XMODEM, known as the *CRC option*, has been devised that has a two-byte figure. This figure is called a cyclical redundancy check (CRC-16) and detects errors at least 99.99% of the time. CRCs are explained in Chapter 10.

By convention, the receiver must indicate to the sender that the CRC option is to be used by sending the character C instead of NAK to request start of transmission. Since not all versions of XMODEM incorporate the CRC option, the receiver should switch to sending NAK if, after several C attempts, no response is received. The block format using the CRC option is shown in Table 8.2.

Offset	Contents
0	SOH (start-of-header character: ASCII 01)
1	Block number: starting with 1, but wrapping to 0 after FF
2	One's complement of the block number (255 − block number)
3 to	
130	128 bytes of data
132	CRC high byte
133	CRC low byte

Table 8.2: XMODEM block format with CRC option

YMODEM ENHANCEMENTS

YMODEM is XMODEM with some enhancements. The protocol was devised by Chuck Forsberg of Omen Technology, Inc., and is gaining ground. The features that it adds to XMODEM are listed below.

- CRC-16 error checking, as discussed above.

- Optional 1K blocks. Sending STX (02H) at the start of each block instead of SOH (01H) signifies that the block that follows is to be 1024 bytes instead of 128 bytes. Mixed blocks of 1024 bytes and 128 bytes can be sent in a single transmission.

- CAN-CAN abort. Two consecutive CAN (18H) characters indicate that the file transfer is to be aborted.

- Batch file transmission. Several files can be sent at once. For each file a block numbered zero is sent. This block contains the file name in lowercase, terminated by ASCII 0. The file name can include a path name, in which case it is delimited by a forward slash (/), as is the convention in the UNIX system, not a backslash (\) as in MS-DOS systems.

Block zero can also have any of the following four additional fields:

1. The file size as a decimal string followed by a space. If the file size is sent, the receiving program will know which padding characters in the last block, if any, to disregard.

2. The modification date as an octal number measured in seconds from Jan 1 1970 GMT, with zero if the date is unknown, followed by a space. Using GMT avoids problems that can arise when people in different time zones are working on a file and it is not clear which is the latest version.

3. A file mode as an octal string (UNIX systems only), followed by a space.

4. A serial number as an octal string. The field is set to zero if there is no number.

The rest of the block is set to nulls. Further files are sent, each with its own file name block. A null file name terminates batch transmission.

Note that many host computers, and some networks, cannot handle continuous 1K blocks. However, YMODEM supports software handshaking using XON/XOFF, which, depending on the networks and mainframes being used, may allow the transmission of 1K blocks.

XMODEM ADVANTAGES AND DISADVANTAGES

X MODEM's main advantages are its simplicity and its universality. It does have several disadvantages, however, that result primarily from the fact that it is often used in ways for which it was not originally intended.

It is not capable of solving the word length problem discussed Chapter 7. No special transformations take place to allow eight-bit data to be sent with seven-bit communications parameters. Accordingly, if binary data are being sent, eight-bit communications must be possible. The protocol specifies eight data bits, no parity, and one stop bit, even if only seven-bit text data are being transmitted.

Also, there is no protection against binary data being mis-interpreted as control signals. If the receiving device, for example, always treats ASCII 4 as meaning the end of transmission, you can only send three blocks since block four will contain ASCII 4 as its block number. This means that even if the data to be transferred are in acceptable ASCII format the protocol itself can add unacceptable codes.

There is no requirement for full duplex communications, since the receiving device must wait for transmissions to cease before sending NAK. However, this can result in delays, especially over long-distance links. More sophisticated protocols allow one block to be acknowledged or rejected while a later block is being sent.

SUMMARY

W hile XMODEM has its shortcomings, several of which are addressed in YMODEM, and while more efficient and clever protocols have been and will be implemented, XMODEM retains its popularity due to its universality and ease of implementation.

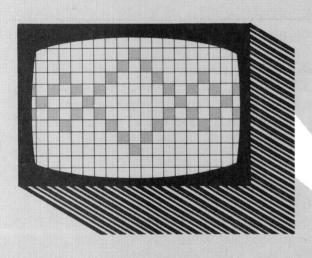

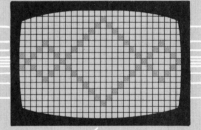

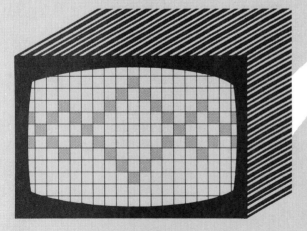

CHAPTER 9

Kermit

INTRODUCTION

*T*he Kermit protocol was developed at Columbia University in New York, primarily to facilitate file transfer between mainframe computers and microcomputers. Kermit is useful for users who want to use a mainframe computer as a storage medium or as an intermediary. With Kermit they can upload programs they have written to the mainframe for subsequent downloading by others, even though the programs are not executable by the mainframe itself.

Kermit has been widely accepted, especially in the academic world, and versions now exist for all major computers. These various versions tend to be circulated from hand to hand. There are both public domain, and copyrighted but freely distributed, Kermit programs that not only include the protocol but are complete programs in themselves offering the communications functions needed for the particular machine on which they are running. Accordingly, Kermit is sometimes used to refer to the protocol and sometimes to a program incorporating it. We will deal only with the protocol here.

Recently, commercial packages have started to offer Kermit as an option. However, Columbia requests that, in keeping with the spirit in which they freely made the protocol available, people incorporating it into commercial products should not charge for doing so. Columbia does maintain a copyright in its own Kermit programs and documentation, so, strictly speaking, it is not in the public domain.

The following information is derived from the *Kermit Protocol Manual,* Fifth Edition, by Frank da Cruz; the Kermit Windowing Protocol Draft Version 1.2; and from a paper on long packets circulated by Frank da Cruz.

USING KERMIT

*T*he normal method of using Kermit to transfer to and from a mainframe is to start with a communications program running on the microcomputer that incorporates both a terminal emulator and the Kermit protocol. Using the micro as a terminal emulator, the user instructs the mainframe to run its own version of Kermit. Then the microcomputer is switched into Kermit mode and the file transfer can proceed. There are optional *server commands* that enable the microcomputer to send a number of commands to the mainframe while still in Kermit mode. However, this facility is not always implemented.

SYSTEM REQUIREMENTS

Kermit makes very few demands on the communications capacities of the two machines it is running on. It can cope with systems limited to seven-bit characters, even when the data to be transmitted are in eight-bit form. No other handshaking is required other than that provided in Kermit itself. Full duplex operation is not required for basic operation.

CHARACTER ENCODING

U nlike XMODEM, Kermit does transform all transmitted characters into standard printable characters (ASCII 32 through 126). This way, nonprintable characters can be transmitted without causing the receiving computer, or intervening communications equipment, to handle them in special ways.

CONTROL CHARACTERS

Control characters are characters 0 through 31, and 127. When these characters are found in the data to be sent, Kermit translates them into printable characters by XORing them with 64 and preceding them by a prefix character (normally #). So Ctrl-A, which is ASCII 1, becomes A, which is ASCII 65, and is transmitted as #A. The receiving program, when it detects the prefix character, discards it and converts the following character back to a control character by XORing it with 64. If the prefix character itself is to be sent, it is sent twice, e.g., ##.

EIGHT BIT TO SEVEN BIT CONVERSION

Kermit can convert characters that must consist of eight bits (i.e. extended characters or binary data) to seven bits. This is necessary when either computer, or the intervening communications path, is not capable of handling eight-bit characters. Kermit checks each character to see whether bit seven is set (i.e., the character is greater than 127). If bit seven is set, the character is preceded by a *quoting* character (normally &). The character itself is then sent as a seven-bit character. Since this adds considerably to the length of the transmission, it should be avoided whenever eight-bit characters can be sent.

KERMIT INFORMATION CHARACTERS

Characters that do not form part of the data being transmitted but that contain numeric information internal to Kermit, such as the length of a packet, are encoded by adding 32 (unless otherwise stated). For example, to indicate a packet length of 26, the number 26 must be sent in the LEN field of the packet. Kermit adds 32 to 26 and sends the ASCII character that corresponds to the total, which in this case is the colon, or ASCII 58.

This function is known as the *tochar( )* function. The corresponding decoding function is the *unchar( )* function. It decodes characters known to represent internal Kermit numeric information by subtracting 32 from all characters passed to it.

The greatest number that the tochar( ) function can encode is 94. This is because 95 + 32 = 127, which is a control character, and any characters greater than 127 require eight bits. Accordingly, 94 is something of a magic number within Kermit because it limits the size of various parameters.

Characters that represent themselves (for example when a device wants to indicate to another device which character it wants to precede control characters) are sent as themselves; no special encoding is necessary.

COMPRESSION

When long sequences of the same character are sent, Kermit can compress them. It sends a sequence consisting of a repeat character, which is normally a tilde (˜), followed by the repeat count, followed by the character to be repeated. Since the repeat count must be stored in a single character encoded with the tochar( ) function, the maximum repeat count is 94. Control characters and eight-bit characters can be repeated; when they have been encoded as two characters (see above) both characters are sent following the repeat count.

PACKETS

*T*he primary message unit used by Kermit is the *packet*. A packet can consist of data or other information such as acknowledgment or error messages. The types of packets are listed below. Each type is designated by a letter of the alphabet that is stored in the type field of the packet.

D	Data packet
Y	Acknowledge (ACK)
N	Negative acknowledge (NAK)
S	Send initiate (exchange parameters)
B	Break transmission (EOT)
F	File header
Z	End of file (EOF)
E	Error
T	Reserved for internal use
I	Initialize (exchange parameters)
A	File attributes
R	Receive initiate

> C   Host command, containing Server commands to the mainframe
>
> K   Kermit command
>
> G   Generic Kermit command

As you can see, Kermit uses these packets to send all sorts of information between communicating computers from initiation signals to end of file information as well as actual data.

GENERAL PACKET FORMAT

The general packet format is illustrated in Table 9.1. The length of individual packets is variable, and packets of different lengths can be sent in one transaction. However, according to the standard protocol, the maximum total packet length is 96 characters. The LEN field, which is the second field in each packet, contains the number of characters following the LEN field up to and including the CHECK field. Thus, LEN would be 94 for the longest permitted packet of 96 bytes. (See the section on Long Packets later in this chapter for exceptions to this rule.)

Offset	Contents	Meaning
0	MARK	Normally ^A to mark start of packet
1	LEN	Number of ASCII characters after this field
2	SEQ	Sequence number, modulo 64, starting at 0
3	TYPE	Packet type (see below)
4 ...	DATA	Contents of the packet, if applicable
[end]	CHECK	Checksum

Table 9.1: Kermit packet format

Each packet has a sequence number, starting with zero for the initialization packet. The number is modulo 64, meaning that it returns to zero after 63.

INITIATING THE TRANSACTION

A transaction is started when the transmitting device sends an initiation packet containing various parameters. This is acknowledged by the receiving device, which in turn indicates, in the ACK packet, which parameters and options it can support. For example, the sending device may be able to handle eight-bit words but does not know whether the receiving device has this capability. The initiation sequence establishes what common ground exists.

In some cases the exchange of information can take place without automatically being followed by a file transfer. In this case the initialization packets are known as *Init-Info packets* and are type I.

INITIALIZATION PACKETS

The packet that the sender transmits to initiate a transaction, and that is normally sent repeatedly until acknowledged, contains six basic characters and a number of additional characters of information in its data field. The acknowledgment packet contains similar information regarding the receiving device. The required fields are listed below.

MAXL	The maximum length packet that can be received
TIME	The number of seconds the other device should wait before reporting a timeout
NPAD	The number of padding characters that should precede each incoming packet
PADC	The control character required for padding
EOL	The character required to terminate an incoming packet, if any
QTCL	The character used to precede control characters (normally #)

The initialization packet can also contain the following optional information:

QBIN   The QBIN character is used to precede characters that have the eighth bit set when the parity bit cannot be used for data.

CHKT   CHKT indicates the check type: 1 means one-character checksum; 2 means two-character checksum; 3 means three-character CRC-CCITT.

REPT   REPT is the prefix used to indicate a repeated character: space (32) means no repeat count processing; tilde (~) is the normal repeat prefix.

CAPAS   CAPAS is a bit mask indicating various KERMIT capabilities. Each bit is set to 1 if the capacity is present. Each character in the field consists of six bits, transformed into a printable character.

QBIN is the character used to precede characters that have the eighth bit set when the parity bit cannot be used for data. It must be different from the QTCL character. A Y response in this field agrees to do eight-bit quoting if requested, N refuses to do eight-bit quoting, and "&" or any other character in the range 33–62 or 96–126, agrees to do eight-bit quoting with the character used. Normally the ampersand is used if eight-bit quoting is to be done (i.e. the program knows it is dealing with a seven-bit channel).

The CAPAS character(s) contains information about the capabilities being requested (if sent by the transmitting device) or authorized (if sent by the receiving device). There will generally only be one CAPAS byte, but more can be added by setting the low order bit of the last byte. Table 9.2 illustrates the format of the first two CAPAS bytes. The third and subsequent characters are available for users.

DATA TRANSFER

Once the initialization packets have been exchanged, the data to be sent are transmitted as a series of data packets. The

A: First character

Field #	Bit	Meaning
1	5	Reserved
2	4	Windowing
3	3	A packets (file attributes)
4	2	Long packets
5	1	Reserved
	0	1 if another CAPAS follows

B: Second character

Field #	Bit	Meaning
6	5	Reserved
7	4	Reserved
8	3	Available for users
9	2	Available for users
10	1	Available for users
	0	1 if another CAPAS follows

Table 9.2: Kermit capability bytes

receiving device must acknowledge each packet with an acknowledgment packet. The transmitting device must wait for each packet to be acknowledged before sending another packet. If the receiving computer reports an error by sending a NAK packet, the sending computer retransmits the packet.

If a NAK packet is received for a packet, this is treated as an equivalent of an acknowledgment of the previous packet. If a packet arrives more than once, the receiver sends an acknowledgment packet and discards the duplicate.

Several files can be sent in one transaction. For each file, Kermit sends a file header packet, one or more data packets, and an end of file packet. Some implementations of Kermit also send a type A packet after the fileheader packet. This packet contains additional information about a file such as its date of creation, size, and type (e.g., ASCII, EBDIC, or binary). When there are no more files to send, the transmitting computer sends an end of transmission packet. The interchange of packets is illustrated in Table 9.3.

Event Name	Sender to Receiver	Receiver to Sender
SEND INIT (send initiate)	Parameter packet	
ACK (acknowledge)		Parameter packet
HEADER (file header)	Header packet	
ACK (acknowledge)		Acknowledge packet
DATA (data packet)	Data packet	
ACK (acknowledge)		Acknowledge packet
(LOOP TO DATA if more data to send)		
EOF (end of file)	EOF Packet	
ACK		Acknowledge packet
(LOOP TO HEADER if more files to send)		
EOT (end of transmission)		
ACK		Acknowledge packet

Table 9.3: Kermit transaction sequence

INTERRUPTING TRANSFER

Kermit has an optional feature that enables file transfer to be interrupted. Sending an EOF packet with a D in the data field means "discard the file." The EOF packet should still be acknowledged. Sending an ACK packet with an X in the data field means "interrupt this file." Sending an ACK packet with Z in the data field means "interrupt the whole batch of files."

A fatal error at either end is indicated by an error packet, which terminates the transaction. This should be sent if the above procedures do not work, since it is possible that the other device does not support the interruption options.

INTERPACKET DATA

Any terminating character required by the system can be added at the end of the packet. This is normally a carriage return. It is not considered part of the package when computing the length or check total. Other characters such as handshaking characters can also be sent between packets.

ERROR CHECKING

*T*here are three ways of computing the CHECK figure, and during initialization the communicating computers must agree upon the method to be used. The receiving program computes the check figure the same way for each packet, and sends a NAK packet if the check figure does not agree with the number in the CHECK field.

Note that the ACK and NAK packets also have check figures and the sending computer should also make sure that the check figures it receives are correct. However, if an ACK or NAK packet has a bad checksum, the sending computer will not send back a NAK. Normally, such a packet is simply ignored and treated as if it was not received.

ONE-CHARACTER CHECKSUM

The simplest CHECK figure to compute is the one-character checksum. It is computed as follows. The ASCII values of all the characters in the block are added up. If the characters are eight-bit, all eight bits are included. Bits 6 and 7 are then added to the number formed by bits 0 through 5. The result is converted into a printable character. Thus if s is the arithmetic sum of the ASCII characters, then

check = tochar(32 + ((*s* + ((*s* AND 192)/64)) AND 63))

TWO-CHARACTER CHECKSUM

For a two-character checksum the arithmetic sum of the characters is computed as a sixteen bit figure, and the low-order twelve bits are sent as two characters converted using the tochar( ) function (bits 6 through 11 followed by bits 0 through 5).

THREE-CHARACTER CRC

A sixteen bit cyclical redundancy check figure is calculated and sent as a sequence of three characters formed by applying the tochar( ) function to bits 12 through 15, bits 6 through 11, and bits 0 through 5 respectively. See Chapter 10 for more information about CRC calculations.

OPTIONAL KERMIT FEATURES

*T*here are two optional additional features that make Kermit even more useful. Unfortunately they are not included in all versions of Kermit, and cannot be used on all computers, but they are becoming more readily available.

LONG PACKET EXTENSION

One optional feature, not included in all versions of Kermit, provides for longer packets than originally specified. A request for long packet authorization is made by the transmitting device setting capability number 4, which is bit 2 of the first CAPAS character in the initialization packet. A receiving program that is able to use long packets will respond by accepting the request.

The transmitting program can also specify the longest packet it can accept as input. This length is indicated in two bytes, MAXL1 and MAXL2. MAXL1 contains the character represented by

$$32 + (length / 95)$$

MAXL2 contains the character represented by

$$32 + (\text{length MOD } 95)$$

MAXL1 and MAXL2 are contained in the second and third bytes of the last CAPAS byte. If they are missing, but the long packets bit was set, the default length of 500 is used. The regular MAXL byte must also be set with a length conforming to the original Kermit protocol, since it may not be known until after initialization whether the receiving computer is capable of implementing long packets.

When long packets have been agreed on, an individual packet is treated as a long packet if its LEN field is set to zero. The format of a long packet is shown in Table 9.4.

Offset	Contents	Meaning
0	MARK	Normally ^A to mark start of packet
1	LEN	Set to zero to indicate a long packet
2	SEQ	Sequence number, modulo 64, starting at 0
3	TYPE	Packet type
4	LENX1	Character of (32 + INT(length / 95))
3	LENX2	Character of (32 + (length % 95))
3	HCHECK	Checksum of LEN, SEQ, TYPE, LENX1, LENX2
4 ...	DATA	Contents of the packet, if applicable
[end]	CHECK	Checksum of all bytes from LEN on

Table 9.4: Kermit long packet format

SLIDING WINDOWS

One complaint about the Kermit protocol is that it is slow when used over long distances because of the need to wait for an acknowledgment after each packet is sent. This can add a considerable overhead, especially when the two computers are at remote locations and other media, such as networks, intervene. In such cases the waiting time can exceed the transmission time.

To solve this problem, the windowing amendment has been proposed. It enables successive packets to be sent without waiting in between for an acknowledgment. Acknowledgment is still expected in due course, however. Not all communications systems can implement sliding windows, since each computer must have full duplex capability.

Requesting Sliding Windows

A request for windows is made by setting bit 4 of the first CAPAS byte in the initiation packet. At the same time the window size (i.e., number of data packets to be sent sequentially) is placed in the first field following the last CAPAS byte. The maximum number of packets in a window is 31. The receiving device then indicates to the sender whether it is capable of using this feature.

Note that long packets and sliding windows are not mutually exclusive. Table 9.5 shows the format of a send init packet when both extensions are being implemented.

Starting the Transfer

The sender transmits the file header packet in the normal way and waits until it has been acknowledged before sending data packets. This is because the receiving device is presumed to be saving the data to a file and cannot do this until the file name has been successfully received as part of the file header packet.

Transferring Data

On receipt of each packet, the receiving device sends an ACK or NAK message as appropriate. It keeps a list of which packets have been successfully received. Likewise, the sending device keeps a list of which packets have been acknowledged by the receiving device.

Since the number of packets in a file can exceed the number of packets allowed in a window, the window is seen as sliding through the file. It moves on when the first packet in the window

Offset	Contents	Meaning
0	MARK	Normally ^A to mark start of packet
1	LEN	Number of ASCII characters after this field
2	SEQ	Sequence number, modulo 64, starting at 0
3	TYPE	Packet type (see below)
4	MAXL	The maximum length packet that can be received
5	TIME	Time-out time
6	NPAD	The number of padding characters
7	PADC	The control character required for padding
8	EOL	Terminating character if any
9	QTCL	The character preceding control characters
10	QBIN	Character preceding characters over 127
11	CHKT	Check type
12	REPT	Repeated character prefix
13	CAPAS	Capability bit mask
14	WINDW	Number of packets in a window
15	MAXL1	High order (MOD 95) of maximum length
16	MAXL2	Low order (MOD 95) of maximum length
17	CHECK	Checksum

Table 9.5: Send init packet with long packets and sliding windows

has been successfully transmitted (i.e., acknowledged by the receiving device). The sliding window concept is illustrated in Figure 9.1.

On receipt of a NAK packet, the packet referred to is retransmitted, unless it is outside the current window in which case the first unacknowledged packet is retransmitted. The latter event occurs if a packet number is damaged.

Bad Checksums

One problem with this system is that if a packet is received with a bad checksum, sending a NAK may confuse matters. The NAK could be indicating damage to the packet number field that caused the bad checksum; in that case the NAK would be for the wrong packet.

Packet type	Packet number	Sent?	ACK Received?		
Init	0	Yes	Yes		
File header	1	Yes	Yes		
Data	2	Yes	Yes		
Data	3	Yes			
Data	4	Yes			
Data	5			Current Window	
Data	6				
Data	7				
Data	8				
Data	9				Limits of Window
Data	10				
Data	11				
Data	12				
Data	13				
Data					
Data	N				
End of file	+ 1				

Figure 9.1: Sliding window concept (sender's point of view)

Kermit solves this problem by having the sending device ignore any ACK or NAK packets that have a bad checksum. If the receiving device receives a data packet with a bad checksum, it should either send a NAK for the oldest unacknowledged packet, if any, or ignore the bad packet; when the next packet comes it will be clear from the packet number whether one has been lost in the meantime, and a NAK can be sent for the missing one.

Ending the Transfer

The sending device will wait for all the packets to be acknowledged before sending an end-of-file packet. It will then ask for that packet to be acknowledged before sending any further files.

Timeouts

If certain packets are lost, it is possible for each device to be waiting for the other. Accordingly, a sending device should respond to a lengthy gap between receipts by sending the oldest unacknowledged packet, and the receiving device should send a NAK for the oldest unacknowledged packet (if any) or for the next packet due. There should probably also be a limit placed on the number of timeout retries of this type so that the attempt to transfer is abandoned if too many problems occur.

Abandoning File Transfer

A sender can terminate transfer by sending an end-of-file packet with D (for discard) in the data field. A receiver can stop the file transfer by placing an X in the data field of an ACK packet. It can stop all file transfers by placing a Z in the data field. The sender will then send an end-of-file packet with the D in the data field and a sequence number of one greater than the number of the ACK that contains the request to stop.

IMPLEMENTING KERMIT

As you have seen, there are a large number of possible Kermit configurations. However, any implementation you use must be able to support the basic, or "classic", Kermit protocol, recognize any requests for features it does not support, and reject such requests. Accordingly, it is possible to incorporate basic Kermit features into a program without having to worry about how to program CRC error detection or how to implement sliding windows.

FURTHER INFORMATION

You can obtain the full Kermit protocol manual (87 pages) and source code for various versions of Kermit from

Kermit Distribution
Columbia University Center for Computing Activities
612 West 115 Street
New York, NY 10025.

A nominal charge is made to cover expenses. I recommend anyone intending to implement Kermit to acquire the official documentation. It does not, at the time of writing, cover the sliding windows and long packets extensions.

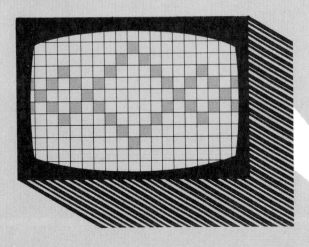

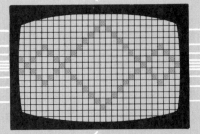

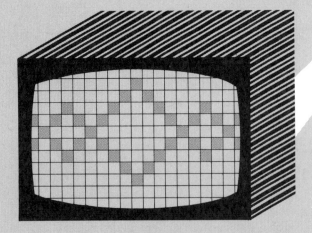

CHAPTER 10

Programming Topics

INTRODUCTION

*T*here are some aspects of communications programming that are of general application and not limited to one machine. I will discuss some of them in this chapter. One of these general subjects is interrupt-driven I/O, since the principles are of fairly general application, and because an understanding is necessary for the more advanced chapters of Part Two. I will also include an introduction to UARTs, a note on Cyclical Redundancy Checks, and a note on circular buffers.

UARTS

A *UART,* or Universal Asynchronous Receiver and Transmitter, is a chip specially designed to handle asynchronous communications. A *USART,* or Universal Synchronous Asynchronous Receiver and Transmitter, is designed to handle both synchronous and asynchronous communications. The following information applies equally to UARTs and USARTs.

THE JOB OF THE UART

The UART has four main jobs:

1. It converts parallel signals coming from the computer's central processing unit into serial signals for transmission out of the computer, and converts serial signals coming into the computer into parallel form for processing by the computer.

2. It adds the necessary start, stop, and parity bits to each character to be transmitted, and strips off those bits from received characters.

3. It ensures that individual bits are sent out at the appropriate baud rate, computes the parity bit on transmitted and received characters, and reports any detected errors.

4. It sets up the appropriate hardware handshaking signals, and reports on the status of incoming handshaking circuits.

CONNECTIONS TO THE UART

Connections to the UART typically consist of the following:

- Eight pins for parallel data transfer
- Two pins for received and transmitted data
- Clock signal from which baud rates are computed
- Received and transmitted handshaking lines
- Control circuits through which the UART can receive instructions and report status
- Interrupt line(s) through which the UART can alert the CPU of a change in state
- A chip select pin that notifies the chip when to act

The UART is not directly connected to the serial socket on the computer, since the voltage levels used inside the computer are lower than those used in serial transmissions. Accordingly, appropriate circuitry is provided by the interface manufacturer for modifying the voltage levels. In addition, logic is provided for monitoring the control signals sent along the computer's *bus,* or data highway, and recognizing which signals are being addressed to the UART. The UART and its associated circuitry are often incorporated on a card known as a serial interface card, which is often an optional accessory for a computer. In other cases, however, the UART comes built into the computer. Printers, modems, and other communications equipment also frequently embody UARTs.

MAIN UART REGISTERS

The UART has several *registers,* which are internal memory locations. These locations typically hold the latest character received; the next character to be transmitted; status information concerning the handshaking signals currently being detected; and information as to whether the chip is ready to receive another character for transmission and whether a character is ready to be sent.

When communications software is receiving data through the UART the first thing it does is read the status register and find out whether a character has been received. If it has, the character itself is read. This reading normally resets the status register so that it indicates that a character is no longer available. When, in due course, another character has been received, the status register indicates this, and the new character can be read.

Transmission is similar. The status register indicates whether the UART is ready to receive a character for transmission. As each character is sent, the status register is returned to its previous state until a new character is loaded into the UART for transmission.

POLLING VERSUS INTERRUPTS

This cycle of examining status, reading, examining status and so on, is known as *polling*. Most UARTs offer an alternative to this method of cycling. They can be programmed to send a special signal known as an *interrupt* when a communications event occurs. The computer must have been programmed to recognize the interrupt and react accordingly.

Interrupts are typically generated when a character has been received or transmitted or when the handshaking signals change. There may be different interrupts for different events, or just one interrupt. In the latter case it is necessary for the computer to examine a special register to find out what caused the latest interrupt. Sometimes the UART can be programmed to generate interrupts on certain events and not others.

Polling

Using the polling approach has two distinct disadvantages. Imagine that your telephone did not have a bell. You would have to keep picking it up every few seconds to see if anyone was there. If you had several telephones on your desk, you would have to keep picking up each one in turn. This is what happens in polling.

The first problem with this method is the processing power that is wasted by a computer that continually scans the status byte. If the computer is dedicated only to handling the specific communications task in hand then this is not a problem. However, computers are often carrying out several tasks at the same time. This applies to minicomputers, and increasingly to microcomputers with the introduction of co-resident programs and concurrent operating systems. In these cases it is desirable to use a method that avoids unnecessary polling.

The second problem with the polling approach is that a new character may be lost while the first is being processed. This depends on two factors: the speed at which characters are being received and the speed at which they are being processed. Typically, incoming characters are displayed on a screen, saved in a file, and printed out. Sometimes they need to be interpreted, as in the case of escape sequences controlling an intelligent terminal. These processes all take time. When a typical microcomputer uses the polling technique, incoming characters can easily be lost if the rate exceeds 1200 baud, even if disk access is not required.

It might be possible to use handshaking controls, either software or hardware, to stop the flow of incoming characters after each character has been received and restart it after it has been processed, but this method is grossly inefficient and might not even work because of the time delay between sending the stop signal and the remote device actually stopping.

Despite these drawbacks, polling is very commonly used in serial communications, and if you are using strictly DOS commands, polling is the only method available on the IBM PC. It is because of these drawbacks mentioned above that few serious communications programs written for the IBM PC stick to the DOS functions.

Hardware Interrupts

Using interrupts is the equivalent of equipping your telephone with a bell. You no longer have to keep picking up the receiver to see if anyone is there. You can wait for the telephone to ring and concentrate on other tasks until that happens. In other words, a UART programmed to send interrupts sends a signal when a relevant event occurs. That way the receiving computer knows when something has happened and doesn't have to waste time checking.

Often a computer has more than one interrupt line, but only one is assigned to a particular chip. This means that although the UART may be programmed to generate an interrupt when one of several different events occurs, the CPU only knows which UART generated the interrupt, not which event within the UART caused it. In other words, you would hear your telephone ring but you would not necessarily know who was there.

Although the use of interrupt driven I/O can save a lot of processing time, it is not without its pitfalls. It is important to remember that an interrupt can be received at any time during the processing cycle. The computer and any software running on it must be designed so that any task can be suspended when an interrupt occurs and resume after the event causing the interrupt has been dealt with.

Remember that a computer may be handling interrupts from two serial cards, the keyboard, a clock, a mouse, and a hard disk, all of which can arrive at any time. As you can see, it is important to be very careful when implementing interrupt-driven programming.

CIRCULAR BUFFERS

Another communications programming topic that can be discussed in general terms is the circular buffer. As we saw in Chapter 3, a buffer is an area of memory used as a temporary holding tank for data awaiting transmission, or received data awaiting processing. One common type of buffer is the *circular buffer*. Characters are retrieved from the buffer on a FIFO (first in/first out) basis. If the buffer becomes full, new characters will overwrite the oldest characters in the buffer first; however, properly implemented handshaking should prevent characters from being received when the buffer is full.

The same method can be used for both input and output buffers. In the case of input buffers under interrupt-driven communications, the characters are placed in the buffer by the interrupt-handling routine. They are then retrieved from the buffer by a routine in the main program loop. A general method of programming for a circular buffer is discussed below.

CREATING THE BUFFER

First, an area of memory is allocated and a variable recording its size (SIZE) is set up. Then variables recording the count (COUNT), and the offset (OFFSET) from the start (STARTPOS) are allocated and initialized to zero.

ADDING CHARACTERS

The offset into the buffer for the new character is calculated as follows:

```
OFFSET = COUNT + STARTPOS
If OFFSET = SIZE, OFFSET = OFFSET – SIZE
```

The character is then placed at offset OFFSET into the buffer. Remember that the first location is offset zero. Thus if the buffer is 128 bytes, the highest offset will be 127, not 128.

If COUNT is equal to SIZE, we have more characters than fit in the buffer, and we increment STARTPOS. If STARTPOS, as a result, points beyond the end of the buffer, it is returned to the beginning. If COUNT is less than SIZE, we increment COUNT. Figure 10.1 shows the stages of filling a circular buffer.

RETRIEVING CHARACTERS

If COUNT is zero, there are no characters available, so the function returns with a FALSE result. The first character to be

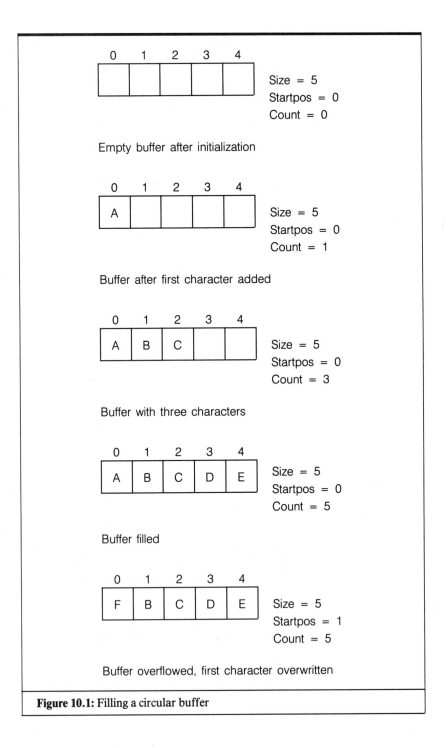

Figure 10.1: Filling a circular buffer

removed is at offset STARTPOS into the buffer. Having taken the character, the function increments STARTPOS (returning to zero when STARTPOS equals SIZE) and decrements COUNT. Figure 10.2 shows the stages of emptying a circular buffer.

```
        0   1   2   3   4
        A   B   C   D   E    Size = 5
                             Startpos = 0
                             Count = 5


Buffer Full

        0   1   2   3   4
       [A]  B   C   D   E    Size = 5
                             Startpos = 1
                             Count = 4


One character retrieved.
The first character is still there but will be overwritten by the
next new character.
```

Figure 10.2: Emptying a circular buffer

CYCLICAL REDUNDANCY CHECKS

Cyclical redundancy checks (CRCs) are a form of error checking. The purpose of an error code is to compute a number that is related mathematically to the string of data corresponding to a message being transmitted. This is done by the transmitting device. The receiving device also calculates the number using the same formula based on the data received. It is hoped that any errors in the data will affect the number computed, so that errors can be detected.

The simplest way to compute a number based on a string of text is simply to add up the ASCII values of all the characters. This is known as a checksum.

The CRC calculation has two main advantages over a checksum. First, it reveals a higher proportion of errors. Second, it is bit-oriented rather than character-oriented, which allows it to work with protocols that produce a stream of bits rather than a stream of bytes.

I have yet to see an explanation of CRC calculations that did not appear to asssume a degree in advanced mathematics. Since neither I nor most of my readers are advanced mathematicians, I will try to explain the calculations in simple terms.

There are several different CRC calculations. In each case, the stream of bits comprising the message is treated as one huge binary number. First, a number (n) of zeros are added to the end of the huge number, to multiply it by 2^n. Then, the number (now even larger) is divided by a number (d) that varies depending on which CRC standard is being used. Only the remainder is retained. This number is transmitted to the receiving device, which then performs the same calculation to make sure the two numbers agree and that no data have been lost or corrupted.

Different communications programs use different versions of CRC. XMODEM, with the CRC option, uses CRC-CCITT. So does Kermit. SDLC uses a modified version of CRC-CCITT. Bisynch uses CRC-16 with EBCDIC and CRC-12 with Six Bit Transcode. With ASCII text, Bisynch uses different methods (vertical/longitudinal checks).

Each CRC performs calculations slightly differently. In the case of CRC-16,

$$n = 16$$

and

$$d = 1100000000000101$$

In the case of CRC-CCITT,

$$n = 16$$

and

$$d = 1100000000000101$$

In the case of CRC-CCITT,

$$n = 12$$

and

$$d = 1100000001111$$

The number (d) that is divided into the huge number is known as a *generator polynomial.* The polynomial is expressed in the following form (using CRC-CCITT as an example):

$$X^{16} + X^{12} + X^5 + 1$$

This expression shows which bits are set in the number used to divide the huge number.

This description explains how CRCs are calculated. However, learning how to program your equipment so that it performs CRC checks is a complicated matter. Some advanced communications equipment already incorporates hardware specially designed to do CRC calculations. For more information, read *Technical Aspects of Data Communications* by John E. McNamara. Some individual examples are given in Part Three of this book.

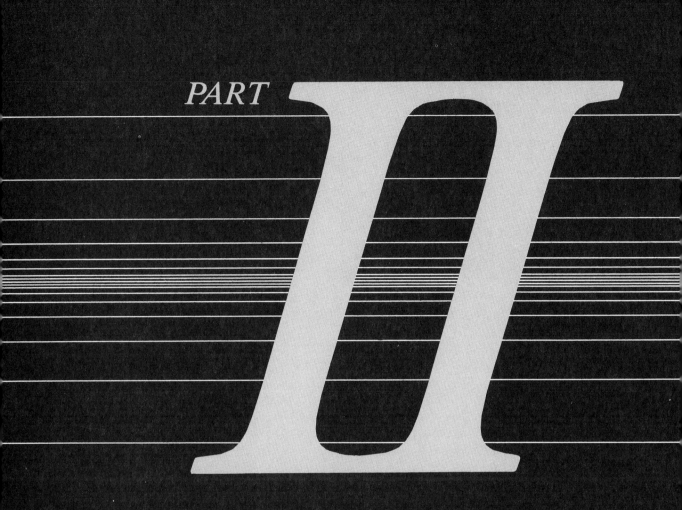

PART

II

**THE IBM PC
AND PC DOS**

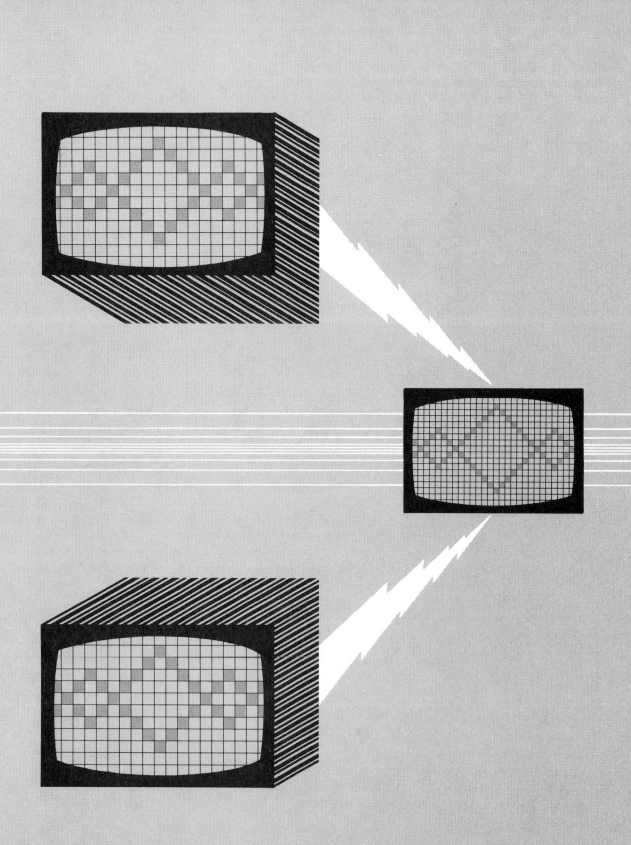

CHAPTER 11

IBM PC Communications
at the User Level

INTRODUCTION

*P*art Two of this book relates to serial communications on the IBM PC. There are four chapters, in increasing order of complexity. This organization enables readers who do not need to program the IBM PC at an advanced level to stop where they feel appropriate.

This chapter describes the operation of serial communications on the IBM PC from a user's point of view. Chapter 12 describes programming for serial communications using the functions available through DOS and BIOS. Chapter 13 covers the 8250 UART chip commonly found in IBM PC systems. Chapter 14 describes the architecture of the IBM PC, focusing on serial I/O, and systems level programming.

HARDWARE

*T*here are several machines in the IBM PC series. There are also a large number of *PC compatible* computers made by companies other than IBM but intended to run software written for the IBM PC. Unfortunately for us, noncompatibilities show up more often in the area of communications than anything else. If a machine is not a *true compatible,* a computer that runs all IBM PC software, there is a good chance that it will not run communications software written for the IBM PC. I will use the term PC to refer to all the machines in the IBM PC series and the true IBM PC compatibles, unless otherwise stated.

THE MAKEUP OF THE COMPUTER

The IBM PC series is designed in a modular fashion, enabling a user to configure a system to his own requirements. The basic PC system unit consists of a circuit board with processor and memory chips, power supply, expansion slots, case, and keyboard. Usually either one or two floppy-disk drives and/or a hard disk are included. The IBM PC itself does not come with a serial interface built in, although some of the PC compatible computers do.

The PC has a number of expansion slots inside it, in which various devices can be added. It is simple to add an expansion card: one simply has to remove the cover of the machine and insert the new card into a free slot. Sometimes small switches on the card or within the IBM PC have to be set; this is explained in the instructions that come with the card. Expansion cards perform such functions as printer interfacing, control of a monitor, and serial I/O.

SERIAL CARDS

There are three main types of cards that provide asynchronous serial communications capability:

1. Dedicated interface cards such as the IBM Asynchronous Communications Adapter

2. Multifunction cards that incorporate a serial interface with other features, such as the AST Six-Pack Plus

3. Internal modems, such as the Hayes Smartmodem 1200B, that fit on a card inside the PC and are seen by the PC as serial interface cards.

The IBM PC-AT enhanced model comes with an adapter card that has both serial and parallel interfaces.

There are literally dozens of different expander cards that you can buy; they must, however, adhere to the specifications of the standard IBM serial interface so that software written for the standard interface can run on them. Very few people use cards that only function as serial adapters. There is a limited number of expansion slots in a PC, and it is more economical in terms of slots to use a multifunction card that typically includes one or two serial ports, a parallel port, a clock, and expansion memory.

Internal modems tend to be slightly cheaper than external ones, but have two main disadvantages: they use up a slot, and they lack the status lights on the front panel. In addition, they can only be used for one series of computers, which is a problem if you ever plan to change computers or even use another type temporarily. However, if you use an external modem you do need to connect it to a serial card or multifunction card.

Synchronous communications cards are also available, but you see them much less frequently than asynchronous cards; they are used almost exclusively in corporate environments for communications with mainframes.

Installation

The IBM PC is designed to handle up to two serial devices, which are referred to as *COM1* and *COM2*. Most cards with serial interfaces have a switch to designate them as either COM1 or COM2, and if you have two serial cards you should designate one as each. Likewise, you can configure most software so that it directs output to either COM1 or COM2 as desired.

Interrupt Lines

When installing serial devices, you will often encounter problems that require an understanding of interrupt lines. The instruction manuals very rarely cover the problem adequately. Certain circuits are provided in the PC that carry signals from devices to indicate certain events. These circuits are numbered IRQ0 through IRQ7 on the IBM PC and IRQ0 through IRQ15 on the IBM PC-AT. Normally, IRQ4 is assigned to the primary serial interface, and IRQ3 to the secondary serial interface. However, certain manufacturers of peripheral devices assume that only one serial device will be used, and use IRQ3 for other purposes, sometimes without mentioning the fact.

If, after installing two internal serial devices such as an internal modem and a multifunction card, you find that one of them does not work, and you have checked that you have set one up as COM1 and the other as COM2, you should investigate which interrupts are being used by which devices. I have had this problem twice myself, and in neither case was given any assistance from the documentation.

Sometimes the problem involves the clock on a multifunction card. Often the clock is set up to generate an interrupt that is unnecessary except for special programming that requires the multifunction card clock rather than the system clock. On one multifunction card I came across there was a jumper to select which interrupt was to be generated by the clock. I had to remove the jumper altogether because all the interrupts available were being used by one or another device in my system.

I have also had problems with a Microsoft mouse, which was attached to an internal interface card. The documentation did not even mention that there was a jumper to select which IRQ line was to be used. Oddly enough it was in the documentation for TopView supplied by IBM that I discovered that there was a jumper that could be changed on Microsoft's mouse.

Fortunately, there are often some spare IRQ lines. Line 2 on the IBM PC (not on the AT) is usually spare (although labeled 'Reserved' in the *DOS Technical Reference Manual*), and line 5 on the PC can be used if you do not have a hard disk.

If you have a PC with a bus version of a mouse, an internal modem, a multifunction card, and a hard disk (not an unlikely combination), you will have to set the mouse to use IRQ2.

There are ways of adding more than two serial interfaces, but special adapters must be used, and since there is no standard means of doing this you should refer to the instructions accompanying such a device if you need to use one.

All these problems make one wonder sometimes whether the Apple designers deserve all of the criticism they received for designing closed systems (the Apple IIc and the Macintosh) without expansion slots. It can sometimes take an expert a considerable amount of time to resolve incompatibilities between different peripherals competing for the same addresses and interrupt lines.

OVERVIEW OF DOS

*I*n order to use your PC you must have a decent grasp on the working of DOS, the disk operating system. An operating system consists of one or more programs designed to govern the basic operation of a computer. PC-DOS is the operating system most commonly used on the IBM-PC, although others such as CP/M and Xenix are available. PC-DOS can also be used on some non-IBM computers that are true compatibles. MS-DOS is very similar to PC-DOS and is used on non-IBM computers that are not quite true IBM compatibles. I will refer from hereon to DOS, and most of what I say in this chapter and the following one will be relevant both to PC-DOS and to MS-DOS. Although DOS stands for disk operating system, you will see that it covers more than just disk operations.

I will not attempt a complete description of DOS here. There are several excellent books on the subject, as well as the manual that comes with DOS, and the *DOS Technical Reference Manual.* I will concentrate here on the aspects of DOS that are relevant to serial communications at the user's level.

BOOTING DOS

DOS is contained in the COMMAND.COM file and two hidden files that are on a disk but cannot be seen in a directory of the disk. In the case of IBM PCs, these two files are IBM-BIO.COM and IBMDOS.COM.

Before the computer can be used, the operating system must be *booted,* or loaded into memory. If the computer has a hard disk, it automatically loads the operating system when you switch on the machine. If you only have floppy-disk drives, you must place a special bootable disk in drive A that contains the three files COMMAND.COM, IBMBIO.COM, and IBMDOS.COM.

WHAT DOS DOES

DOS can be divided into two parts: DOS commands and DOS functions.

DOS Commands

The DOS commands take care of certain housekeeping tasks such as copying disks, reading a disk directory to see what is on it, and setting the date and time. For DOS to perform these tasks the user must enter the necessary commands at the keyboard. Some of the tasks are built into DOS, and others are extra programs contained in program files that must be present when the command is issued. For example, the command DIR, to read a directory, is recognized by DOS when it is typed at the keyboard, but the command DISKCOPY is contained in the DISKCOPY-.COM file that must be on a disk in the computer before it can be run.

DOS Functions

The second part of DOS consists of functions such as opening and closing a file, and writing a character on the screen that cannot be called from the keyboard but can only be called through programs. These are known as DOS functions. These functions make it easier to write programs. Because they are used a lot, it is convenient to have them resident in memory at all times, so that many different programs can use them.

DOS DEVICE NAMES

DOS has a set of names of devices built into it. The device names are listed in Table 11.1. The colon after the names is optional. You can add other device names by installing device drivers, discussed below.

AUX:	The first serial/parallel adapter port
COM1:	The first serial/parallel adapter port
COM2:	The second serial/parallel adapter port
CON:	The console (keyboard and screen)
PRN:	The first parallel printer
LPT1:	The first parallel printer
LPT2:	The second parallel printer
LPT3:	The third parallel printer
NUL:	A nonexistent (dummy) device

Table 11.1: DOS device names

DOS is configured so as to allow up to two serial interface adapters, or two different devices at the same time, and they are referred to in the documentation and by the operating system as COM1 and COM2. Each adapter card generally has a switch configuring it as COM1 or COM2, so that it knows which messages are for it and which are for the other adapter. There are also two sets of addresses reserved for these devices, as is explained in Chapter 14. As mentioned above, an internal modem is seen as a serial interface and is also addressed as COM1 or COM2.

DEVICE DRIVERS

You can extend and modify DOS by using additional files known as *device drivers* that normally can be recognized by the extension .SYS in the file names. These device drivers contain extensions to the operating system that allow it to control particular physical or logical devices. They are loaded into memory at the same time as DOS and effectively become part of it.

If a file with the name CONFIG.SYS is contained in the root directory of the boot disk, DOS reads that file when it is booted and follows the commands contained in it. These commands consist of instructions in the following form:

DEVICE = DEV.SYS

where DEV is the name of the particular device driver to be loaded.

THE MODE COMMAND

One of the DOS commands that you can use is the DOS command MODE. It has two purposes.

1. It assigns a device name and tells DOS whether the standard printer is on one of the parallel ports or one of the serial ports.

2. It configures the parameters of a serial port.

ASSIGNING DEVICES

LPT1 (line printer 1) is the device name used for the standard printer. By default, LPT1 refers to a printer connected to a parallel port. If you want it to refer to a printer attached to a serial port (i.e., to assign LPT1 to COM1 or COM2), you type

    MODE LPT1: = COM1:

or

    MODE LPT1: = COM2:

This way you can change the output of a program without having to rewrite the program. For example, if a program tells DOS to send something to LPT1, the output will go to whichever device you told DOS was LPT1. The program does not need to know that you have your modem set up as COM1 and a serial printer as COM2 so you want printed output to go to COM1. If, however, the program tells DOS to send something to COM1, it will go there regardless of what you have assigned to COM1 with the MODE command.

A program can also ignore DOS altogether and access the device directly, provided it knows the address of the device and the correct way to manipulate it. This is described in Chapter 14.

SETTING COMMUNICATIONS OPTIONS

MODE is also used to configure the following characteristics of a serial port:

    baud rate
    parity
    data bits
    stop bits
    printer handshaking

This facility is mainly used to configure the serial ports for use with serial printers. When you are using a modem, your communications software will usually ask you for the parameters and set them for you; you will not have to use MODE.

If you aren't using a modem, however, setting your parameters with MODE saves other software from having to set them. The software can tell DOS to print something without having to be concerned with parity and stop bits, or even whether the printer is serial or parallel.

To use the MODE command to set the parameters of the serial port COM1, for 9600 baud, no parity, eight data bits, and one stop bit, you would type

```
MODE COM1:9600,N,8,1
```

Note that the parameters are specified with abbreviations, and are separated by commas.

Baud Rate

The valid baud rates are 110, 150, 300, 600, 1200, 2400, 4800, and 9600. Only the first two characters of each number are required; thus you could set 9600 baud by just entering 96.

Parity

Parity is set by entering N for none, O for odd, or E for even. If you select none of these, even will be set by default.

Data Bits

The number of data bits can be either seven or eight. Seven is the default number.

Stop Bits

There can be either one or two stop bits. If the baud rate is 110, the default is two stop bits, otherwise it is one stop bit.

Printer Handshaking

As you learned in Chapter 1, because the IBM PC is generally configured as a DTE device, it expects to receive handshaking signals on pins 6 and 8. In fact, both of these lines must be high before it will transmit, unless it has been programmed at the systems level. For almost all programs that use the printer, lines 6 and 8 must be high.

Supposing you want to transmit something, but the handshaking lines are not high. If the device to which you are transmitting is a printer, several things could be wrong. The printer could be switched off, or in local mode, or out of paper. But it has most likely turned off the handshaking signals because its buffer is full. DOS should keep trying, and eventually the buffer will clear and more data can be sent.

The final parameter that you can enter at the end of the list takes care of this situation. It is an optional p. This is used when the device in question is a printer. If there is a p at the end of the parameter list, DOS will keep retrying to send a character until you press Ctrl-Break instead of reporting an error when no handshaking signal is detected after a given time period (i.e., a timeout error). This is because printers are typically in a busy state, and therefore trying to suppress communications, for longer periods than other serial devices.

If you do not add the p parameter, DOS will report a Write Error when it is asked to send a character and the handshaking signal is not present. It will then ask you to Abort, Retry, or Ignore the error.

You may ask what you do when you want DOS to ignore handshaking altogether. The answer is that this is impossible. You cannot send anything to a serial port through DOS unless the handshaking lines are high. But in earlier chapters I have described several methods of communicating, some of them using hardware handshaking, some software, and some none at all. In order to use these alternative communications methods, and still go through DOS, you will have to fool DOS into thinking that the handshaking signal is there. If your device supplies a signal on one line but not the other, you can join the two together. If no handshaking signals are supplied, you may be able to feed the computer's own outgoing handshaking signal on line 20 back into lines 6 and 8.

Defaults

Leaving out parameters, but leaving in the commas used to separate them, tells DOS to use the default parameters. Thus to set 9600 baud, even parity, seven data bits, one stop bit, and wait for handshaking, enter

96,,,,p

STANDARD INPUT/OUTPUT

*A*nother feature of DOS allows programs to take their input from a standard input and send their output to standard output. Standard input and output often refer to the keyboard and the video screen respectively but can also refer to other input and output devices. Not all programs use these features; some take their input directly from the keyboard and send it directly to the screen.

To understand this concept better, let us suppose that you want to write to the president of IBM. You could either address your letter to "The President, IBM Corporation" or, if you knew his name, to "Mr J. Smith, IBM Corporation." Now supposing there were a board room coup at IBM between your sending the letter and its arriving. The way you addressed the letter would determine whether the letter would go to the new president or the old president. In a similar way, a program can send its output either to the standard output device—whatever that may be at the time—or to a particular output device.

In the case of programs that use the standard input and output feature, you can use DOS to change the standard input and output devices when you invoke the program. You do this by using the symbols <, >, and >>.

A good example is the DOS SORT command. Suppose that you want to sort a file called NAMES.TXT. To do so, you enter

SORT < NAMES.TXT

SORT would take its input from NAMES.TXT and send it to the default standard output device, i.e., the screen. If you wanted the output printed out on a printer set up as LPT1, enter

SORT < NAMES.TXT > LPT1:

To sort the file and save the sorted data in a new file, enter

SORT < NAMES.TXT > NEWFILE.TXT

To sort the file and add the sorted data to the end of an existing file, enter

SORT < NAMES.TXT >> OLDFILE.TXT

As you can see, standard input and output can be either a physical device or a file. The standard output from a program can even be redirected to become the standard input of another program. To do this you use the symbol ¦. To produce a sorted directory, enter

DIR ¦ SORT

The sorted directory will appear on the screen. The concept of passing output from one program to another is called *piping*. A program that takes its input from standard input, modifies it, and sends the results to standard output is known as a *filter.*

Combining the two concepts, you could print out a sorted directory by entering

DIR ¦ SORT > LPT1:

THE COPY COMMAND

A ny DOS user will be familiar with using the DOS program COPY to copy files. With this command, you can copy a file from a disk in drive A to a disk in drive B by entering

COPY A:NAMES.TXT B:

You can copy a file to a new file with a different name by entering

COPY NAMES.TXT NEWFILE.TXT

In addition to copying files to disks, you can also copy them to devices. To display a file on the screen, you can enter

COPY NAMES.TXT CON:

To print out a file on a printer set up as LPT1:, you can enter

COPY NAMES.TXT LPT1:

You can also use a device as a source for the COPY command. You can write to a disk file from the keyboard by entering

COPY CON: NAMES.TXT

When you are finished, type Ctrl-Z, or press F6, and close the file. You can type directly to a printer by entering

COPY CON: LPT1:

COPY can, to a limited degree, be used to access the serial ports. You can copy a file called THISFILE to COM1, for example, by typing

COPY THISFILE COM1:

You can also instruct a modem to dial a telephone number by placing the appropriate commands in a file and using the COPY command to copy the file to the serial port where the modem is attached. Remember to use MODE to set the parameters first.

You can also control the modem from the keyboard by typing

COPY CON: COM1:

Nothing will be sent, however, until you press F6 or Ctrl-Z at the end of the message. Nor will you see the responses echoed back from the modem.

It would be very useful if you could use the COPY command to transfer a file from a serial port: in other words, to download a file. With the CP/M operating system, on which DOS was based, you can do this, using the CP/M equivalent PIP. How-ever, this feature was not implemented in DOS and its absence apparently resulted in a certain amount of confusion because the DOS manual is inconsistent on the subject. I think the authors of DOS intended to implement the download feature but then changed their minds. You can see this in the manual for PC-DOS 3.1. On page 7-61 it says "You can also use COPY to transfer data between any of the system devices." But on page 7-62 it says "You cannot use COPY to transfer a file using the COM or AUX serial ports." Even this latter statement is incorrect, since you can use COPY to transfer a file *to* a serial port, but not *from* one.

Another error in the DOS manual is on page 7-66. It gives several examples of using COPY, including "copy aux con." This is consistent with the earlier statement that you can transfer data between any two devices, and also consistent with PIP under CP/M, but it is just not true for DOS!

THE CTTY COMMAND

Another DOS command that is useful in serial communications is the CTTY command. With it, you can control the IBM PC from an external terminal. First, connect the terminal to a serial port on the IBM. Then, use the MODE command to set the communications options (baud rate, etc.) on the PC. Finally, type

    CTTY COM1:

for serial port 1, or

    CTTY COM2:

for serial port 2.

From now on, the PC will disregard anything typed at the keyboard, and will take its input only from the remote terminal. Output will be sent to the terminal also, and the input will be echoed back to the terminal.

In order to return the PC to local control, you have to enter

CTTY CON:

at the remote terminal.

It is possible to have remote access to a PC by leaving it switched on and in the CTTY mode and connected to a modem that has been set up to answer the telephone. You can then dial into your PC from a remote PC or another terminal, and give commands to the PC, run programs, and so on.

Only programs that take their input from standard input and direct their output to standard output could be used with CTTY in this way. This includes most DOS commands but excludes most commercial programs.

Incidentally, I experimented with CTTY, and I found that (at least with DOS 3.1) entering a backspace from a remote terminal caused the system to crash consistently. I have not found this problem referred to in any books on DOS, but when I posed the question on CompuServe two other people confirmed my findings.

In addition to the CTTY command, there are other, more powerful commercial programs, including one from Microstuf called Remote, that enable the IBM PC to be run from a remote terminal or another computer over a modem.

COMMUNICATIONS SOFTWARE

*M*ost users of the IBM PC do not write their own communications software; they use one of several standard packages that are available. Probably the most widely used is Smartcom II, from Hayes, which comes free with Hayes modems. Another popular package is Crosstalk XVI, from Microstuf. There is also a program called PC-TALK, which is available under the "user supported software" concept. This means that you are free to copy the program and pass it on, but if you acquire it this way you are invited to submit a payment to the developer on an honor system. This makes you a "registered user," entitled to support and updates.

Some programs, including Crosstalk XVI, offer *script files.* These files, created with a text editor or word processor, contain a series of commands that are followed by the program. For example, you could create a script file that dials up your electronic mail box, gives your password, downloads any waiting messages into a file, and logs off. Another program offering script files is the sophisticated Pro-Yam, from Omar Technology, designed for advanced users. The use of script files is particularly advantageous when expensive information sources are being accessed. A regular user of Dialog, for example, where connection charges can exceed a dollar a minute, would want to speed up the process as much as possible.

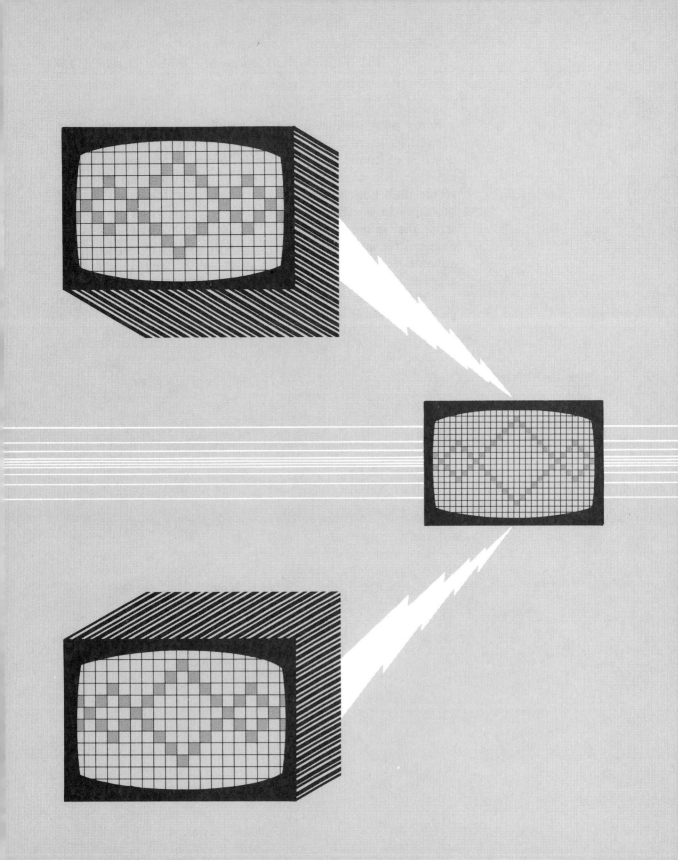

CHAPTER 12

IBM PC Communications
at the
DOS and BIOS Level

INTRODUCTION

*D*OS and BIOS provide a number of built in functions that can be called on by programs and several of these functions relate to serial communications. This chapter describes the DOS functions relating to serial communications, and the advantages and disadvantages of using them.

Programs do not have to use the available DOS and BIOS functions. It is always possible, and often quicker, to achieve the same results in different ways. Relatively few commercial programs use the DOS functions when writing to a monochrome screen, for example, because it is much quicker to move text directly to an area known as the screen buffer. Not even BASIC, which is supplied by IBM with the PC, uses the DOS screen handling functions. With serial communications programming it is often possible to bypass DOS and control the serial ports directly. (These methods are discussed in Chapter 14.) Nevertheless, it is important to be aware of how you can use DOS and BIOS functions in communications programming because you will often find that they provide better compatibility and less programming effort than lower-level programming.

DOS AND BIOS

*T*here are two sets of functions provided with PC-DOS. Some of the functions are stored on ROM chips provided with the machine. These are known as ROM-BIOS functions but are referred to here simply as BIOS. Others are loaded into memory when the operating system is booted. These are known as DOS functions. From a practical point of view they work the same way. There is some overlap between the two, and one is generally recommended to use the DOS functions where a choice is available. However, sometimes the BIOS equivalents are more powerful.

Portability is more likely to be achieved by using the DOS functions, since it is likely that they are implemented in the same way on the various PC-DOS and MS-DOS machines; this is not always the case with BIOS functions.

There are two DOS functions and four BIOS functions that are used in serial communications. These functions are accessed through software interrupts (not to be confused with hardware interrupts, which are discussed in considerable detail in the following chapters).

SOFTWARE INTERRUPTS

Software interrupts are a means of instructing the processor to branch to a different location so that the instructions stored at that location can be carried out. When a software interrupt command is read by the PC's control processor, the processor reads an area of memory known as the *interrupt vector table*. This area contains addresses for each of the possible interrupts. Each address is four bytes long. So, to execute interrupt 21H, for example, the processor looks at offset (4 × 21H) into the interrupt vector table. It then executes the function stored at that address, which ends with an IRET or interrupt return statement. The processor then continues with the next statement in the original program.

In order to access the DOS and BIOS functions, the program places the necessary parameters into the registers, or internal memory locations, of the CPU chip and then triggers the relevant interrupt with an INT instruction. Each language accesses these functions in a different way and I will discuss various languages in Part Three of this book. For the meantime, I will assume that you have some means of loading the registers, calling an interrupt, and accessing the values returned in the registers by the interrupt.

THE DOS FUNCTIONS

The two DOS functions covering serial communications are accessed by issuing an interrupt 21H. They both apply to the AUX device, or the "standard auxiliary device" as it is called in the DOS manual. It is very difficult to ascertain exactly what the auxiliary device is supposed to be. The *DOS Technical Reference*

Manual states "Auxiliary (AUX, COM1, COM2) support is unbuffered and noninterrupt driven." This seems to imply that AUX can be assigned to COM1 or COM2, but there is nothing in either the DOS manual or the *DOS Technical Reference Manual* concerning this. In fact the only reference to AUX in the DOS manual seems to be as a reserved word! The only reference to AUX in the *DOS Technical Reference Manual* is in relation to the serial functions described here. I have also looked in vain for explanations in standard books on DOS and PC programming.

This confusion, coupled with the fact that the *DOS Technical Reference Manual* virtually tells you that the functions are fairly useless, makes me wonder why the functions are there at all. Nevertheless, I have discovered, by experimentation, that the functions do in fact receive and send characters through COM1. If anyone can shed any light on this subject I would be interested. I suspect that it is leftover from CP/M, but that the CP/M functions have only been partly implemented.

The first DOS serial communications function is used for serial input. Serial input is achieved by setting the AH register to 3 and issuing an INT 21H instruction. DOS waits for a character to be received from COM1 and returns it in register AL.

Serial output is the other DOS function and it is achieved by setting the AH register to 4, placing the character to be sent in register DL and issuing an INT 21H instruction. The character will be sent to COM1.

There is no way to set the communications parameters (baud rate, etc.) through the DOS functions. Either the programmer sets BIOS functions or the user has to set the parameters with the MODE command.

Neither DOS serial input nor output returns any error information. These functions are so primitive that they are not very useful. IBM admits in the *DOS Technical Reference Manual* that "For greater control, it is recommended that you use the ROM BIOS routine (interrupt 14H) or write an AUX device drivers (sic) and use IOCTL." IOCTL is a DOS function for handling device drivers.

Interestingly, there is no information provided in the *DOS Technical Reference Manual* about how to use BIOS functions, or even where to find that information. The answer is in the technical reference manuals for the hardware, which incorporate the source code for the BIOS.

THE BIOS FUNCTIONS

To use the four BIOS functions relating to serial communications you must access them through the BIOS interrupt 14H. You place a number from 0 to 3 in AH indicating which of the four functions you require. You then place a port number in the DX register. The port number is zero for COM1 and one for COM2.

The Set Communications Parameters Function

The first BIOS function, function 0, is used to set communications parameters. It is accessed by setting the AH register to 0 and the DX register to the port number, placing a byte representing the parameters in the AL register, and issuing INT 14H. Bits 0 and 1 of the parameter byte govern the word length. For eight-bit words, both bits are 1. For seven-bit words, bit 1 is 1 and bit 0 is 0. Bit 2 indicates the number of stop bits. A value of 0 represents one stop bit, and 1 represents two stop bits. Bits 3 and 4 indicate parity, as shown in Table 12.1. Bits 5 through 7 indicate baud rate, as shown in Table 12.2.

Table 12.3 shows an example of the parameters for the typical setting of 1200 baud, eight data bits, no stop bits, and no parity.

Bit 4	Bit 3	Parity
0 or 1	0	None
0	1	Odd
1	1	Even

Table 12.1: BIOS interrupt 14H parity settings

When the parameters have been set, the function returns the current port status in the AX register. See the Get Port Status section below.

Bit 7	Bit 6	Bit 5	Baud
0	0	0	110
0	0	1	150
0	1	0	300
0	1	1	600
1	0	0	1200
1	0	1	2400
1	1	0	4800
1	1	1	9600

Table 12.2: BIOS interrupt 14H baud rate settings

Bit	Setting	Value	Meaning
7	1	128	
6	0		1200 baud
5	0		
4	0		No
3	0		parity
2	0		One stop bit
1	1	2	Eight
0	1	1	data bits
Byte computation		131	

Table 12.3: Example of communications parameter byte

The Transmit Character Function

The next BIOS function, function 1, is used to transmit characters. It is accessed by setting AH to 1 and DX to the port number, placing the character to be sent in register AL, and issuing INT 14H. Note that the character will not be sent until the incoming handshaking lines are high. Even with the relatively low-level BIOS functions, hardware handshaking cannot be overridden.

Under normal programming practice you would call the Get Port Status function (see below) and only call the Transmit Character function when you know that the conditions are favorable (i.e., incoming handshaking lines are high).

On return, register AH will indicate any error conditions found. If bit 7 of AH is 0, then the transmission was successful. If bit 7 of AH is 1, the remaining bits indicate which error occurred, using the same coding as described under Get Port Status below.

The Receive Character Function

The Receive Character function, function 2, is accessed by setting AH to 2, DX to the port number, and issuing INT 14H. The BIOS waits until a character is received from the serial port or a timeout has expired. When a character is received, it is placed in AL, and any error condition is reported in AH.

If AH is 0, then no error occurred. If it is not 0, then bits 0 through 7 indicate the error condition as described under the Get Port Status function. However, if bit 7 is set indicating a timeout error, the remaining bits are not predictable.

Normal programming practice is not to call Receive Character repeatedly, but to call Get Port Status repeatedly and only call Receive Character when it is known that a character is available. This gives you much more control, since you can choose your own timeout periods and can be doing something useful between receiving characters rather than just waiting for the next one.

This method also avoids problems arising from a bug in the BIOS of early IBM PCs, which incorrectly reported a timeout as a parity error.

The Get Port Status Function

The Get Port Status function, function 3, is accessed by setting AH to 3, DX to the port number, and issuing INT 14H. It provides various information about the current status of the serial port and returns the current port status in register AX. Table 12.4 shows what the various bits mean.

AH register bit	Meaning if set
7	Timeout error
6	Transmitter shift register empty
5	Transmitter holding register empty
4	Break detect
3	Framing error
2	Parity error
1	Overrun error
0	Data ready

AL register bit	Meaning if set
7	Received line signal detect
6	Ring indicator
5	Data set ready
4	Clear to send
3	Delta receive line signal detect
2	Trailing edge ring indicator
1	Delta data set ready
0	Delta clear to send

Table 12.4: Interrupt 14 port status codes

Delta means that the relevant signal has changed since the last time the port status was read. For example, bit 5 of the byte returned in the AL register indicates whether the received DSR handshaking signal is high or not. Bit 1 indicates whether the state of the DSR signal has changed since the last time the port status was read. The delta information is generally used in connection with interrupt-driven I/O, and since this is not possible through DOS and BIOS the delta bits are not really applicable here.

HANDSHAKING UNDER BIOS

B IOS behaves very strangely regarding handshaking signals. The Set Parameters function does not set up any handshaking signals. The Receive Character function turns the DTR signal on and turns the RQS signal off (!) and then waits for a character to be received, returning a timeout error if none is received within a certain time period. The Transmit Character function sets DTR and RQS, and waits for both DSR and CTS to be set by the other device. If they are not set, the function returns after a timeout.

So the strange thing is that while BIOS insists on *receiving* two handshaking signals it only *provides* one to the other device. Not only that, but it actually *turns off* RQS when waiting to receive! So if the device with which you are communicating is another PC or another device that insists on receiving two handshaking signals you will have to fake the missing signal by joining RQS to DTR at the remote device end of the cable and disconnecting RQS at the PC end.

Remember that until you have called the Send Character or Receive Character functions, your outgoing handshaking signal is low, or off. If the other device expects a handshaking signal, you must call Send Character or Receive Character first. From then on, at least DTR will remain high.

However, following the normal procedure of calling Get Port Status to see if there is a received character and calling Receive Character if there is one can cause problems. You might never receive anything because you do not call Receive Character until something is there, and nothing will be there because the handshaking line you are putting out is low because you have not yet called Receive Character!

I recommend that as part of the initialization sequence, after setting the parameters you call the Receive function once even if you know there is nothing there, just so that you can turn on DTR. And don't be fooled by the Transmit Character function setting RQS and think you do not have to jumper the wires together, because the Receive Character function will turn RQS off again just when you want it.

Also, note that there is no way to turn off the handshaking signals when you have finished communicating. So do not rely on the modem disconnecting when you close down your program or you may have a few large telephone bills.

DOS AND BIOS VERSUS DIRECT CONTROL OF THE HARDWARE

When you are working with serial communications you must weigh the pros and cons of using the DOS and BIOS functions or actually programming the hardware yourself.

THE ADVANTAGES OF DOS AND BIOS

There are three main advantages of using the DOS and BIOS functions.

1. First, they may save programming effort and space in the program file. If the problems have already been solved, why reinvent the wheel? And if the functions are already in memory waiting to be called on, why duplicate them?

2. Second, you can achieve more compatibility with other DOS machines. Some machines, for example, run MS-DOS but are considerably different from the IBM PC in their architecture. They may have different chips and different locations. In fact it is communications more than almost anything else that tends to show up compatibility problems. An attempt to write directly to the hardware on these machines might fail because the hardware is different or in

different locations. However, the DOS functions are almost always the same on all MS-DOS machines.

3. Finally, programs using DOS functions exclusively should be compatible with future versions of the PC and with new environments such as TopView. IBM encourages people to do everything through DOS, because they want to be sure that if they introduce a new machine with different hardware all the existing programs will run on it.

THE DISADVANTAGES OF DOS AND BIOS

Unfortunately, in the case of serial communications there are limitations on how much can be achieved through DOS and BIOS. Not everything that can be done through systems level programming can be achieved by sticking to the DOS and BIOS functions.

Interrupt-Driven I/O

The 8250 UART chip used in IBM PC compatible serial boards can be programmed to generate interrupts when certain events, such as the receipt of a character, occur. The advantages of this method, and the techniques involved, are more fully discussed in the following chapters. You must, of course, program the chip directly; the DOS functions cannot help. This method is almost essential for sophisticated communications software running at the higher baud rates.

Handshaking

All the DOS functions require two incoming handshaking lines to be high before they will transmit. In some cases, such as direct connection to a large computer, hardware handshaking is not provided. However, it is usually possible to fake an incoming handshaking signal by connecting the outgoing handshaking lines to the incoming ones. Only one outgoing handshaking line is provided. By programming the UART directly, you can control the handshaking lines and/or ignore them as you wish.

Break

You will recall that a Break is a method of interrupting some mainframe programs. It is not possible to transmit a Break using only DOS and BIOS functions. It can only be done by programming the UART directly, as explained in the next two chapters.

CONCLUSION

*N*either the DOS nor the BIOS communications functions are very adequate, and they have clearly been given a low priority by the designers. Perhaps we can expect better support in future versions of DOS, but in the meantime it will be necessary to use lower level programming in order to make full use of the serial communications capabilities of the PC. The next two chapters explain how to do this.

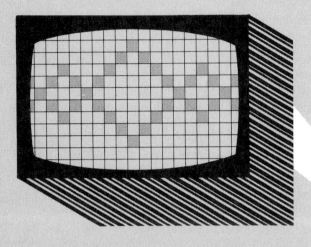

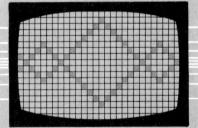

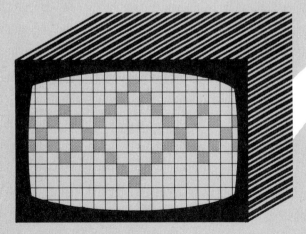

CHAPTER 13

The INS 8250 UART

INTRODUCTION

We discussed UARTs in general in Chapter 10. In this chapter we take a detailed look at the INS 8250 chip, the UART almost always used in the IBM PC for asynchronous serial communication. If you plan to do systems level communications programming on the IBM PC you must have an understanding of this chapter. You can also use this chapter if you want to program other computers that use the same UART.

REGISTERS OF THE UART

Like the CPU chip, the UART contains registers, or internal memory locations. There are three types of registers:

1. Control registers, which receive commands from the CPU

2. Status registers, which are used to inform the CPU of what is going on in the UART

3. Buffer registers, which hold characters pending transmission or processing

The way that the registers are accessed depends on the architecture of the computer in which the 8250 is installed. In the case of the IBM PC, values to be placed in the registers are sent to an appropriate I/O address by means of an OUT command to the CPU chip. Registers to be read are accessed by means of an IN instruction accompanied by the appropriate address. The actual addresses are given in the next chapter.

CONTROL REGISTERS

There are four control registers that are used to receive commands from the CPU.

Line Control Register

The line control register is used to set the communications parameters. The meanings of each bit in the register are shown in Table 13.1 and a more detailed discussion of each bit follows.

Bit	Meaning
0	Word length: least significant bit
1	Word length: most significant bit
2	Stop bits
3	Parity enable
4	Parity select
5	Parity one
6	Break
7	Divisor latch access bit (DLAB)

Table 13.1: The bits of the line control register

- Bits 0 and 1 record the word length. (The meanings are shown in Table 13.2.)

Bit 0	Bit 1	Word length
0	0	5
0	1	6
1	0	7
1	1	8

Table 13.2: Word Length Select in Line Control Register

- Bit 2 records the number of stop bits. If bit 2 is zero, one stop bit is used. If bit 2 is one, two stop bits are used unless the word length is five bits, in which case one and a half stop bits are used.

- Bit 3 enables parity: if it is zero, no parity bit is transmitted or expected. If it is one, a parity bit *is* generated or expected.

- Bit 4 selects even parity if it is set to one, and odd parity if it is set to zero. Bit 4 is ignored unless bit 3 is also set.

- Bit 5 makes the parity bit a logical zero if set. If bit 5 is zero, the parity bit becomes a logical one.

- Bit 6 is used to generate a break command. It forces output to the spacing condition (logical zero) until bit 6 is set to zero.

- Bit 7 is called the divisor latch access bit (DLAB). If it is set to one, a read or write operation accesses the divisor latches of the baud rate generator (see below). If it is set to zero, read and write operations access the receiver and transmitter buffers or the interrupt enable register.

Modem Control Register

The modem control register controls the handshaking signals sent out from the UART. Each bit of the register is listed in Table 13.3 and explained below.

Bit	Abbreviation	Name
0	DTR	Data terminal ready
1	RTS	Request to send
2	Out1	User defined output 1
3	Out2	User defined output 2
4	Loop	Test mode loop-back

Table 13.3: Modem control register bits

- Bit 0 is used to set the Data Terminal Ready output (DTR) to logical zero (i.e., enable a remote device to transmit to us). If bit 0 is zero, DTR is set to logical one (i.e., request remote device not to send to us).

- Bit 1 is used in exactly the same way to control the Request to Send (RQS) output.

- Bits 2 and 3 control auxiliary user-defined outputs known as OUT1 and OUT2. These do not correspond to anything on serial cards I have seen. They can be used to control outgoing Carrier Detect and Ring Indicator, but do not generally seem to be wired up that way. On the IBM PC, bit 3 must be set to enable interrupt-driven I/O, due to an anomaly in the design.

- Bit 4 enables diagnostic testing mode.

- Bits 5 through 7 are permanently set to zero.

Interrupt Enable Register

We discussed interrupt driven I/O in Chapter 10. It is possible to instruct the 8250 to generate an interrupt signal whenever certain events occur. The interrupt enable register is used to tell the 8250 which events should cause interrupts.

If your program uses the polling method, no interrupts are enabled. Instead, the program continually examines the status registers to see whether anything has happened. The bits corresponding to each interrupt are shown in Table 13.4.

Bit	Interrupt Set
0	Data available
1	Transmitter holding register empty
2	Receiver line status
3	Modem status
4–7	(Always zero)

Table 13.4: The bits of the interrupt enable register

Baud Rate Divisor Latches

The baud rate is set by recording in two registers a number by which the clock input (1.8432 MHz) must be divided. The resulting frequency is 16 times the baud rate. These two registers are the divisor latch least significant byte (DLL) and the divisor latch most significant byte (DLM). The divisors used to generate different baud rates are shown in Table 13.5.

Note that you are not limited to the conventional range of baud rates. Intermediate values can also be generated by selecting an appropriate divisor. This can be useful for applications such as process control.

Baud rate	Decimal	Hex	LSB	MSB
300	384	180	1	80
1200	96	60	0	60
2400	48	30	0	30
4800	24	18	0	18
9600	12	0C	0	0C

Table 13.5: Baud rate divisors for common baud rates

STATUS REGISTERS

The three different status registers report to the CPU what is happening in different parts of the UART.

The Line Status Register

The line status register is used to obtain information concerning the receipt and transmission of data. The meanings of the individual bits are illustrated in Table 13.6 and are elaborated on below.

Bit	Abbreviation	Name	Meaning if set (i.e., 1)
0	DR	Data ready	An incoming character has been received and placed in the receiver buffer register.
1	OE	Overrun error	A character has been received before the last one was removed for processing.
2	PE	Parity error	An incoming character's parity bit is wrong.
3	FE	Framing error	A received character does not have a valid stop bit.
4	BI	Break interrupt	A break is being received: i.e., the received data input line is zero for more than the maximum character length.
5	THRE	Transmitter holding register empty	UART is ready to receive a new character to transmit.
6	TSRE	Transmitter shift register empty	TSR is waiting for a character from the THR.
7	[SPARE]		This bit is permanently set to zero.

Table 13.6: Line status register bits

- *Data ready* means that a character has been received from outside. This bit remains set until the character has been read from the receiver buffer register.
- *Overrun error* means that a character was received before the previous character had been read. This indicates that characters are being received faster than they are being processed.

- *Framing error* means that a valid stop bit was not detected following the latest character received.

- *Break interrupt* means that a break signal was received.

- *Transmitter holding register empty* means that the UART is ready to receive a character for transmission.

- *Transmitter shift register empty* means that the UART is not in the process of transmitting a character. This register is used in the parallel to serial conversion process, and its status is not normally tested by communications software.

Modem Status Register

The modem status register gives information about the status of the handshaking lines. The meanings of its individual bits are listed in Table 13.7.

Bit	*Name*	*Meaning if set*
0	Delta CTS	Clear to send line changed.
1	Delta DSR	Data set ready line changed.
2	TERI	Trailing edge ring indicator: RI has changed from on to off.
3	Delta RLSD	Received line signal detect changed.
4	CTS	Clear to send input is high (OK).
5	DSR	Data set ready input is high (OK)
6	RI	Ring indicator is high.
7	RLSD	Received line signal detect is high.

Table 13.7: The bits of the modem status register

Bits 1, 2, and 4 are known as *delta bits*. They indicate a change since the last time the register was read. This is used in interrupt-driven programming. You will recall that the interrupt enable register can be set so as to generate an interrupt when the modem status changes. However, several events can cause this change. When this type of interrupt is detected, a program usually examines the delta bits to see what has changed. It then acts accordingly (by suspending transmissions while the incoming handshaking line is low, for example). Bit 2 also indicates a change, but is only set when the corresponding line (line indicator) changes from 1 to 0. Bits 1, 3, and 4 are set when the voltage on the corresponding lines changes in *either* direction.

The second four bits give the actual status of the individual lines: i.e., whether they are high or low, rather than whether they have changed. A value of one means that the line is high, enabling communication. A value of zero means that the line is low, inhibiting communication.

It is not necessary for incoming handshaking lines to be high in order for the 8250 to transmit. It is up to the programmer to test for their status if he or she wants to; hardware handshaking can be ignored if appropriate. Remember that the BIOS transmit function will not transmit unless both lines are high; this is because it specifically tests for this status.

Interrupt Identification Register

The interrupt identification register provides information about the current status of pending interrupts. Bit 0 is set to one if there is no pending interrupt. If it is set to zero, bits 1 and 2 indicate which interrupt is pending according to the chart in Table 13.8. Bits 3 to 7 are always zero.

Bit 2	Bit 1	Interrupt pending
1	1	Line status
1	0	Received data available
0	1	Transmitter holding register empty
0	0	Modem status

Table 13.8: Interrupt identification register bits

BUFFER REGISTERS

The third category of registers in the UART are the buffer registers. There are two buffer registers: receiving and transmitting.

Receiver Buffer Register

The receiver buffer register holds the last character received. Once it has been read, the line status register indicates that the receiver buffer is empty until another character is received. If the second character is received before the first character has been read, an overrun error is reported.

Transmitter Holding Register

The transmitter holding register holds the next character to be transmitted. The character is placed there by your program. The line status register indicates when the character has been transmitted.

PROGRAMMING THE 8250: POLLING METHODS

*I*f you want to program your 8250 to use the polling method and not interrupts, your program must start by setting the interrupt enable register to zero. If transmitting, the program must then keep running through a loop as long as there are characters to send. It repeatedly reads the line status register and the modem status register. When the transmitter buffer register is empty and the appropriate handshaking signals are high it sends the next character.

Remember that the UART does not handle any software handshaking. If you are using XON/XOFF, for example, you also have to test for a character received that might be an XOFF character, and then wait for an XON to be received before resuming transmission.

The receiving process is similar to the transmitting process: you must first set up any handshaking signals needed by the other device, and then loop, testing the line status register for a character received, and reading the character from the receiver buffer register if a character is indicated.

With full duplex communications—when you are transmitting and receiving at the same time—your program must test both for received characters and for readiness to transmit. It should also test for keyboard input as part of the loop.

PROGRAMMING THE 8250: INTERRUPT METHOD

*T*o program a computer to use interrupt-driven I/O, you first have to install an appropriate section of code known as an interrupt handler. There is an example of an interrupt handler in Chapter 18. With this code installed, the computer branches to your interrupt handler whenever it receives an interrupt from the 8250. In the case of the IBM PC, the address to which to branch is set in the interrupt vector table. Next, you would set the appropriate bits of the interrupt enable register of the 8250.

When the interrupt handler receives an interrupt it initially only knows that the 8250 caused the interrupt, nothing else. It first examines the interrupt identification register to see what caused the interrupt. This might indicate Data Ready, meaning that a character has been received, in which case it is read from the receiver buffer register. Usually, the character is then placed in a buffer where it can subsequently be accessed by your main program (i.e., not the interrupt handler).

The interrupt could also indicate that the transmitter holding register is empty, in which case the next character to be sent, if any, is placed in the transmitter holding register.

If the interrupt was caused by another event, then the delta bits of the line status register or modem status register, as appropriate, must be examined to see what has happened, and appropriate action taken.

Note that with interrupt-driven programming there is usually a regular program running at the same time as the interrupt handlers. The interrupt handlers and the main program can communicate by leaving characters for each other to pick up in the receive and transmit buffers that are set up in memory. These are normally circular buffers (as described in Chapter 10).

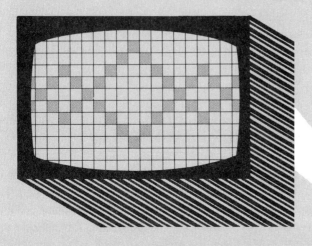

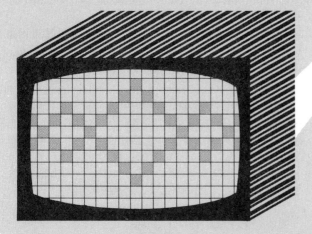

CHAPTER 14

The IBM PC at the Systems Level

INTRODUCTION

*T*his is the final, and most advanced, chapter of Part Two. I will start with a brief overview of the 8088 processor chip and the general architecture of the IBM PC, and then discuss serial communications at the systems level. Because of the complexity of the subject, a basic knowledge of computer operation concepts is assumed.

IBM PC ARCHITECTURE

*I*t is important to understand the main components and design of the IBM PC, so that you will be able to see where serial interface adapters fit in.

THE 8088 PROCESSOR

The central processor chip of the IBM PC is the 8088. Some compatibles use the 8086, and the IBM PC-AT uses the 80286. All these chips are from the Intel family, and the following remarks apply to all of them. I will use the term CPU to refer to whichever processor is in use.

There are four main 16-bit registers, known as AX, BX, CX, and DX. Each of these registers can also be seen as two eight-bit registers. These are AH, AL, BH, BL, CH, CL, DH, and DL. Loading the low-order byte of a 16-bit number into AL and the high-order byte into AH is the same as loading the whole number into AX, and so on with BX, CX, and DX.

The largest registers can only hold 16-bit numbers. Since a 16-bit number can only address 64K of memory, memory must be identified by placing numbers into two registers. These two numbers are known as the *segment* and the *offset* of the address. The segment indicates an address starting at a 16-byte boundary within memory. An eight-bit segment register can identify one of 64K segments each starting on a 16-byte boundary: in other words, within a total memory space of one megabyte. The offset indicates a number of bytes following the start of the segment.

Since the IBM PC's total memory capacity is 640K, any address within memory can be identified by the use of an eight-bit segment register plus a 16-bit offset into that segment. The segment registers are CS (code segment), SS (stack segment), DS (data segment), and ES (extra segment).

An offset can be held in one of the registers AX, BX, CX, and DX, or in one of the registers specifically intended for offsets. These are IP (instruction pointer) for the offset into the code segment, SP (stack pointer) and BP (base pointer) for offsets into the stack segment, and SI and DI (source index and destination index) for offsets into the data segment.

BUS ARCHITECTURE

A *bus* is a collection of circuits connecting devices within a computer. The IBM PC has three buses:

1. The *data bus*. This bus consists of an eight-bit path along which data travel in parallel.

2. The *address bus*. This bus is a 20-bit wide path along which addresses are transmitted.

3. The *control bus*. This bus consists of a number of circuits for controlling devices: generally, instructions as to what to do with the information found on the data bus and the control bus.

ACCESSING MAIN MEMORY

When the processor wants to read memory, it sends the address out along the address bus, and sets a signal on the read line of the control bus. The relevant memory device recognizes the address and puts the contents of that address onto the data bus for return to the processor.

The process for sending data to memory is the reverse of this procedure. A voltage is asserted on the write line of the control bus, the address is placed on the address bus, and the data are placed on the data bus.

I/O ADDRESSES

In addition to up to 640K of main memory that is accessed in terms of segments and offsets, it is also possible to access memory known as I/O memory. 768 I/O addresses, or *ports* as they are often called, are available, and a typical device uses several of these.

You can access these I/O addresses by sending IN and OUT instructions to the processor. For example, to send the contents of the AL register to port 3F8H, the assembly language command is

    OUT 3F8H, AL

To read from the port, whose address is in the DX register, and put the result into the AL register, the command is

    IN AL, DX

Various languages offer different ways of effecting IN and OUT commands, and these methods will be discussed later under the individual languages.

When the CPU receives one of the above commands, it sends a signal along the I/O read command line or the I/O write command line as appropriate. When an I/O device detects a signal on one of these lines, it must first check the address bus to see whether the instruction is for it. Since the highest I/O address is 300H, only the lowest nine bits of the address bus need be connected to I/O devices.

As with main memory, an I/O device is capable of placing data on the data bus when a read request is received, and taking data from the data bus when a write request is received. However, an I/O device may well do more than just place data on the data bus when it receives a read request. It may take other appropriate action: for example, when the 8250 UART receives a read request addressed to its receiver buffer, it not only sends out the contents of the receiver buffer but also clears out the buffer so that the same data will not be read again the next time. It may also reset any interrupts it generated when the buffer became full.

INTERRUPTS

We have already discussed the difference between hardware and software interrupts. Because we are talking here about programming at the systems level we will only be concerned with hardware interrupts in this chapter. A number of interrupt lines exist in the IBM PC and IBM PC-AT. These consist of actual circuits, and interrupts are generated by raising the voltage level on the appropriate circuit and maintaining that level until the interrupt has been reset, or *acknowledged*.

There are 8 IRQ (interrupt request) lines in the IBM PC and 16 in the IBM PC-AT. They are shown in Tables 14.1 and 14.2 respectively.

The IRQ lines do not go directly to the CPU chip, but rather to a dedicated interrupt controller chip. This chip is the 8259A Programmable Interrupt Controller (PIC). The IBM PC-AT has

IRQ line	Device
NMI	Nonmaskable interrupt
0	Timer
1	Keyboard
2	Reserved
3	Serial port 2
4	Serial port 1
5	Hard disk
6	Floppy disk
7	Parallel port 1

Table 14.1: IBM PC IRQ lines

IRQ line	Device
NMI	Nonmaskable interrupt
0	Timer
1	Keyboard
2	Gate from controller 1 to 2
3	Serial port 2
4	Serial port 1
5	Parallel port 2
6	Floppy disk
7	Parallel port 1
8	Clock
9	Redirection from IRQ 2
10	Reserved
11	Reserved
12	Reserved
13	Coprocessor
14	Hard disk
15	Reserved

Table 14.2: IBM PC-AT IRQ lines

two of these controllers. The PIC prioritizes the interrupts and prevents the chaos that would arise if the interrupts could arrive at any time and in any order.

The following notes apply to the PIC as it is actually programmed for the IBM PC; it can behave differently in some ways if programmed differently. It is set up this way by the PC's Power On Self Test (POST) routine, which is executed when the machine is switched on.

The priorities in the IBM PC are assigned with the highest to IRQ0 and the lowest to IRQ7. If several devices are all demanding attention by raising their assigned IRQ lines, the controller will have several inputs high. It will pass these on to the CPU in the order of their priorities.

The PIC has a register where interrupts from the various devices are enabled. By default, IRQ3 and IRQ4 are not enabled.

Therefore, if you want to commence interrupt driven communications you have to instruct the PIC to enable the appropriate interrupt line. This is done by reading the register with an IN command from port 21H, setting the appropriate bit off (bit 4 for IRQ4, bit 3 for IRQ3) and writing the value back with an OUT command to port 21H. This is in addition to enabling interrupts on the UART as mentioned in the Chapter 13.

When the PIC receives an interrupt, assuming that the appropriate interrupt has been enabled and no other interrupt is pending, it asserts a positive voltage on the INT line to the CPU. The CPU then acknowledges the interrupt by means of a signal to the PIC, and requests the PIC to indicate which interrupt occurred. The PIC sends a number (via the data bus) to the CPU (this number is 8 plus the IRQ number). In other words, for IRQ4 (primary serial adapter) it sends the number 12 (0CH) and for IRQ3 (secondary serial adapter) it sends the number 11 (0BH). The CPU then executes the appropriate section of code by saving the current program address on the stack and executing a far call to the memory location pointed to by the interrupt vector for that interrupt. Therefore, an interrupt on IRQ4 from the primary serial adapter has a similar effect to executing INT 0CH in software.

If you do not want the CPU to deal with interrupts for a while because it is involved in some critical task that cannot be interrupted, you can instruct it to ignore interrupts by using a CLI (clear interrupt flag) command. Interrupts are restored by means of an STI (set interrupt flag) command. The result of these commands is to change the setting of the interrupt flag (IF) within the CPU.

Setting the IF does not change anything within the PIC, which will still send a signal on the INT line to the CPU when it receives an interrupt on one of the IRQ lines. It merely prevents the CPU from acknowledging the interrupt to the PIC or from taking any action that would otherwise have been taken. However, the INT line will remain high, and the CPU will be able to identify the source of the interrupt and deal appropriately with it once an STI instruction is executed.

Other pending interrupts are not necessarily lost either. The devices maintain their IRQ lines high, and the PIC knows that those devices need attention. Once interrupts have been enabled again, it will pass on any pending interrupts in order of their priority.

The facility to suspend interrupts can, however, result in a loss of information since the device in question won't generate a second interrupt (for example, if a second character is received by a UART) until the first one has been acknowledged.

As you have seen, there are two reasons why an interrupt from a device might not gain the immediate attention of the CPU: it might be conflicting with another device of higher priority, or interrupts might have been suspended. Nevertheless, the use of the interrupt method is normally faster than waiting for a complicated program loop to execute.

I mentioned that the CPU acknowledges the interrupt from the PIC. This acknowledgment instructs the PIC to notify the CPU which interrupt occurred. However, a further acknowledgement is required. The interrupt handling software (i.e., the code jumped to through the interrupt vector table) must itself acknowledge the interrupt to the PIC by sending the value 20H to port 20H. If this does not happen, the PIC will not send any more interrupts on to the CPU.

Figure 14.1 summarizes the interrupt handling logic within the IBM PC. This has been simplified slightly, and the pin connections rearranged for clarity. The complete logic diagram can be found in the technical reference manual for the machine in question.

SERIAL INTERFACE BOARDS

The IBM PC almost always uses the 8250 UART described in the last chapter. The UART is not built into the PC. It is contained on one of a number of optional expansion boards, as described in Chapter 11.

Other components on the boards handle the physical connection with the motherboard of the computer, connections with the outside world via appropriate sockets, and the conversions of the electrical signals between the voltage levels used within the computer to those used in RS-232-C circuits. However, from a programming point of view, you can ignore these other components and assume that one can address the UART directly and that once the UART reports that it has transmitted a character, the character has in fact been sent out of the computer.

Figure 14.2 shows the main logical circuits of a serial interface adapter. The complete diagram can be found in the technical reference manuals. These manuals contain very useful reference sections on the IBM asynchronous adapter and serial/printer adapter, which are equally applicable to other compatible serial cards.

SERIAL ADAPTER INTERRUPTS

Hardware interrupt number 4 is assigned to the primary asynchronous communications adapter. Interrupt 3 is assigned to the secondary asynchronous communications adapter. Accordingly, assuming your serial card or internal modem is set up as the first adapter, any interrupts generated by its UART will be sent along IRQ4.

The internal priorities within the INS 8250 are fixed. Receive Line Status has the highest priority, followed by Received Data Available, then Transmitter Holding Register Empty and finally Modem Status. Remember that the events do not trigger an interrupt at all unless someone has programmed the interrupt enable register to enable them.

Another interrupt will not be generated until the first one has been reset. The Received Line Status interrupt is reset by reading the Line Status Register. The Received Data Available interrupt is reset by reading the Receiver Buffer register. The Transmitter Holding Register Empty interrupt is reset by reading the Interrupt Identification register (if that was the source of the interrupt) or by writing into the Transmitter Holding register. The Modem Status interrupt is reset by reading the Modem Status register.

SERIAL ADAPTER I/O ADDRESSES

The I/O ports used for serial communications in the IBM PC consist of a series starting at 3F8H for the primary adapter and

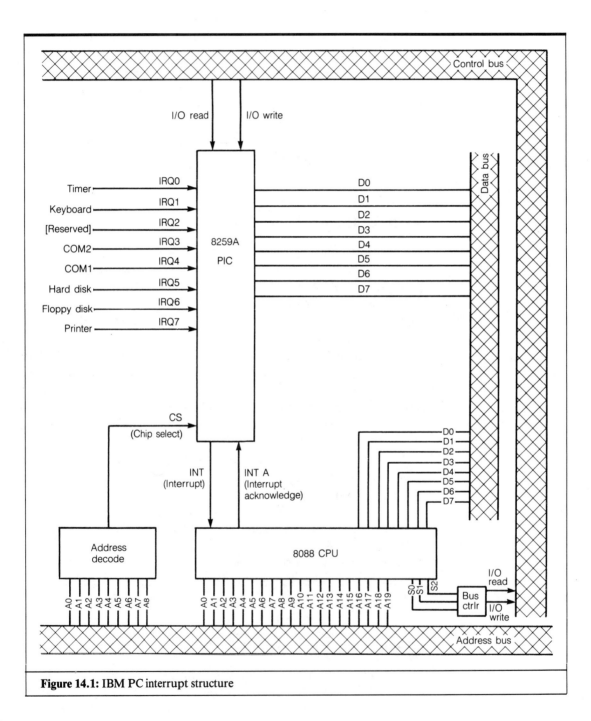

Figure 14.1: IBM PC interrupt structure

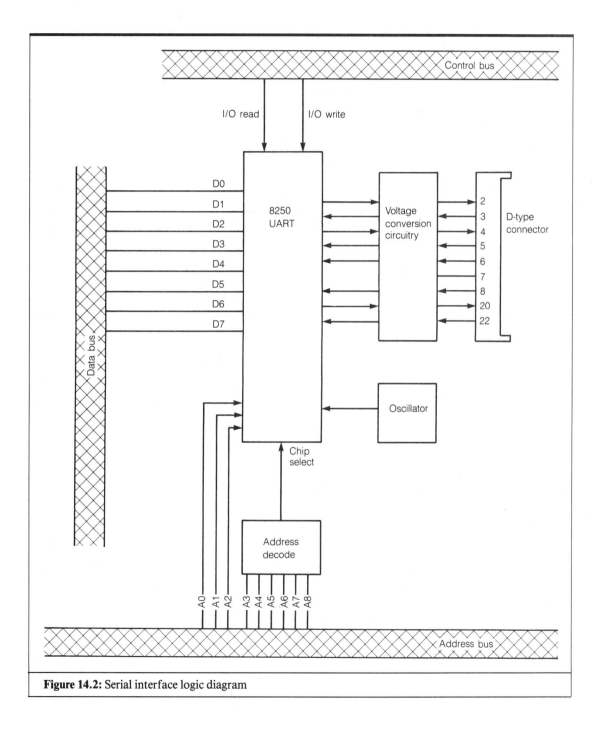

Figure 14.2: Serial interface logic diagram

2F8H for the secondary adapter. To address an individual register within the UART, you add an offset to these base addresses that corresponds to the individual register concerned.

I mentioned earlier that I/O devices only need to be connected to the lower nine bits of the address bus. Both numbers, 3F8H and 2F8H, have the lowest three bits as zero. Accordingly, an offset of up to seven added to one of these numbers does not alter the bits above bit two (counting the lowest bit as zero). The address recognition logic on the serial interface, accordingly, monitors bits three through eight on the address bus. Bits zero through two are connected directly to the UART chip.

When the address recognition logic detects a number in the range of 3F8H through 3FFH, or 2F8H through 2FFH if COM2, it asserts a voltage to the UART on the chip select pin. The UART then looks at the connections to bits 0, 1, and 2 of the address bus, to which it is connected, and to the I/O read and I/O write lines of the control bus, to which it is also connected, in order to decide which action to take. The base addresses and offsets for COM1 and COM2 are shown in Table 14.3.

Notice that some of these registers share the same addresses. Confusion between the transmit and receive buffers is avoided because any OUT command is clearly intended for the transmit

Offset	Primary adapter	Secondary adapter	Register selected
0	3F8	2F8	TX buffer
0	3F8	2F8	RX buffer
0	3F8	2F8	Divisor latch LSB
1	3F9	2F9	Divisor latch MSB
1	3F9	2F9	Interrupt enable register
2	3FA	2FA	Interrupt identification registers
3	3FB	2FB	Line control register
4	3FC	2FC	Modem control register
5	3FD	2FD	Line status register
6	3FE	2FE	Modem status register

Table 14.3: I/O addresses for COM1 and COM2

buffer, and any IN command for the receive buffer. The divisor latches also share addresses with other buffers. These are selected by means of the Divisor Latch Access Bit (DLAB) in the line control register. When DLAB is set to 1, the divisor latches will be addressed. When DLAB is zero, the other registers at the same addresses will be selected.

PROGRAMMING CONSIDERATIONS

*T*he following points are important to consider when program-ming serial communications at the systems level.

POLLING

If you want to write a program that uses the polling method you have to make sure that it will perform repeated IN instruc-tions to the I/O addresses of the appropriate status registers of the 8250, and test for whether a character has been received or another relevant event has occurred. If a character has been received, the program should send a second IN instruction, this time using the address of the receiver buffer register. If a charac-ter has been transmitted, the next character is sent by means of an OUT instruction to the transmitter buffer register.

INTERRUPT SERVICE ROUTINES

The PC maintains an interrupt vector table, which contains one four-byte address for each interrupt. To set up an interrupt vector, you should use DOS interrupt 21H function 25H. You place the address of your interrupt service routine at DS:DX, place the function number 25H in AH, place the interrupt vector to be replaced in AL, and issue INT 21H.

Interrupt 21H function 35H tells you the address currently held in the interrupt vector table for any interrupt. You should use this before changing any interrupts, so that you can store the address of the existing ISR for that interrupt. Before terminating, your program should restore the previous vector.

To find out an ISR address, place the interrupt number in AL, 25H in AH, and issue INT 21H. ES:BX will point to the existing ISR.

Your ISR starts by saving the contents of the registers on the stack. This is because when it has finished its task it must return the processor to the state it was in when the interrupt occurred. Next, the ISR must find out what caused the interrupt, and react accordingly. The ISR must also acknowledge the interrupt to the PIC. Finally, it must restore the contents of the registers and end with an IRET, or interrupt return, command.

It is not possible to write an ISR without using at least some assembly language. If you want to write an ISR, read Chapter 18, which gives further information and examples.

Incidentally, if you have ever wondered how the pop-up programs like SideKick work, it is in a similar way. When they are installed, they replace the vector for the keyboard ISR with a vector pointing to themselves. They then execute a terminate and stay resident DOS function. When you hit a key, a keyboard interrupt is generated, and the processor jumps to the new location containing the pop-up program. The pop-up program then checks to see whether the key that was hit was the one that invokes it. If a different key was hit, the program jumps to the original keyboard ISR.

A NOTE ON THE IBM PC-AT

A word of warning about the AT. The AT is a much faster machine than the PC, and there is a real risk that it can be too fast for the 8250. In other words, it can send instructions to the 8250 faster than the 8250 can process them. For this reason, you should avoid code that sends instructions successively to the 8250; you should place a short jump instruction between successive accesses in order to slow down the stream. This only applies in assembly language programming.

PART

III

SERIAL
COMMUNICATIONS
PROGRAMMING

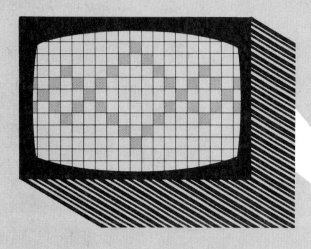

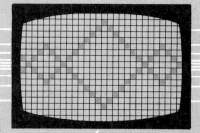

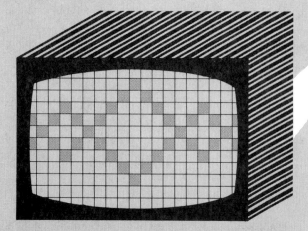

CHAPTER 15

Communicating in BASIC

INTRODUCTION

*T*his chapter and the following two chapters each cover programming for serial commmuncations in one language. The chapters are designed to stand alone, so that most people need only read the chapter for the language they use.

This chapter covers BASIC, specifically advanced BASIC releases 2.0 and 3.0 for the IBM PC written by Microsoft: the versions that are supplied as part of PC-DOS. I am dealing only with interpreted BASIC, and restrict the discussion to the standard commands; in other words, I will not discuss specially coded USR functions.

BASIC has several commands designed for serial communications, and I will discuss each one in turn. These commands are quite powerful, and are much more useful than the DOS and ROM-BIOS functions. It is difficult to interface between BASIC and the operating system, but the power of the BASIC communications functions makes it unnecessary to call the DOS functions directly.

Besides the commands designed specifically for communications, there are also BASIC commands that allow IN and OUT assembly language commands to be issued, enabling direct control, and I will discuss these later in this chapter.

BUFFERING

*B*ASIC automatically sets up a 128 byte output buffer, and an input buffer whose length can be set by the user. Buffers are filled and emptied at an interrupt level. Provided that an input stream has been opened (see below), the receipt of a character causes it to be placed into an input buffer automatically without the program having to do anything, and the successful transmission of a character causes the next character in the output buffer (if there is one) to be selected for transmission. All the program has to do is to request BASIC to send characters, and they will be placed in the output buffer for transmission through the interrupt process. Similarly, BASIC can be requested to input characters, and they will be removed from the input buffer and passed to the program.

It came as something of a personal surprise to me, as a C systems programmer, to discover that the buffer facilities available under BASIC are more powerful than those under C. You cannot write an interrupt handler purely in C because of the lack of an IRET instruction. Accordingly, pure BASIC can run interrupt-driven communications and pure C cannot!

I/O STREAMS

*B*ecause of the similarities between writing to devices and disk files, many of the same commands that are used to access disk files are used to access serial I/O. Although BASIC does not use this term, I will use the expression *I/O stream* to refer to an I/O channel treated as a file. This makes it clear that I am not referring to disk files. An I/O stream can be opened, read from, written to, and closed just like a file on disk.

THE OPEN "COM . . ." COMMAND

An I/O stream is opened by issuing a BASIC command starting with the words OPEN "COM". The remainder of the command depends on the parameters required. The possible parameters are discussed below.

Communications Port

If you want to use COM1, start the command with

OPEN "COM1

If you want to use COM2, start the command with

OPEN "COM2

We will use COM1 in the following examples.

Baud Rate

You can specify any valid baud rate (75, 110, 150, 300, 600, 1200, 2400, 4800, or 9600) after the opening statement. Separate the baud rate from the opening statement with a comma. If you don't specify a baud rate, a default of 300 baud is used. If you want a baud rate of 1200, for example, type

OPEN "COM1, 1200

Parity

The next parameter is a single uppercase character representing the parity option. It can be one of the following:

S	Space
O	Odd
M	Mark
E	Even
N	None

If you don't specify a parity option, Even is selected for you. If you want no parity, your command should read

OPEN "COM1, 1200, N

If eight data bits are used (see below), parity must be N. If four data bits are used, parity cannot be N.

Data Bits

You can specify four through eight data bits by adding the appropriate number to the command. If you don't specify any figure, BASIC selects seven. If you want eight data bits, your command should read

OPEN "COM1, 1200, N, 8

Stop Bits

You can specify either one or two stop bits by adding another number to the command line. The default is one, except when the baud rate is below 300. If you specify four or five data bits and two stop bits, BASIC will generate one and a half stop bits.

Other Communications Options

You can specify various handshaking and other options by adding more parameters to the command. They are listed below:

- **RS** suppresses setting the RTS handshaking signal to the remote device.

- **CS[n]** indicates how many milliseconds to wait for the incoming CTS handshaking signal before returning a Device Timeout error. The number *n* can be from 0 to 65535. If it is zero or omitted, then CTS is ignored. If CS is not added to the command at all, BASIC uses a default of 1000 milliseconds.

- **DS** works exactly the same way as CS, except that it relates to the DSR incoming handshaking line.

- **CD** also works the same way as CS for the incoming CD (Carrier Detect) line. If CD is not added to the command string, Carrier Detect is ignored.

- **LF** causes a line feed to be sent after each carriage return. Whether this is necessary depends on the device with which you are communicating. Printers generally require a line feed after a carriage return.

- **PE** requests parity to be checked. If parity is checked and found to be wrong, BASIC generates a Device I/O error (error 57).

File Number

You must add a file number parameter to the OPEN "COM" command; this number is used in subsequent statements to refer to the I/O stream. For example, to use file number 1 you would add

AS #1

to the command.

Buffer Size

You can specify the size of the input buffer in bytes with an expression of the form

    LEN =

plus a number. If you don't specify a particular size, the buffer will be 128 bytes. The size cannot exceed 256 bytes unless you specified a greater length using the /C: option when you loaded BASIC, and cannot exceed 128 unless you specified a greater buffer length using the /S: option when you loaded BASIC. For example, to select a 1024 byte buffer you type

    BASICA /S:1024/C:1024

followed, of course, by any other command line options required.

Examples

The following example specifies COM1, 1200 baud, no parity, eight data bits, one stop bit, no waiting for incoming handshaking signals, stream number 1, and a buffer of 128 bytes.

    OPEN "COM1:1200,N,8,1,CS0,DS0" AS #1 LEN = 128

The next example specifies COM2, 600 baud, even parity, seven data bits, a 500 millisecond wait for CTS, no wait for DSR, stream number 2, and a buffer of 256 bytes.

    OPEN "COM2:600,E,7,1,CS500,DS0" AS #2 LEN = 256

READING FROM AN I/O STREAM

The BASIC command INPUT # works the same way with I/O streams as it does with files. It reads sufficient data to constitute one variable. For example

    INPUT #1, P$

inputs a string and assigns it to the variable P$.

LINE INPUT # and INPUT$ also work the same way as they do with files. LINE INPUT# inputs a line and allocates it to a string variable. INPUT$ inputs a fixed number of characters. To read five characters from a stream designated as #1, you would enter

    P$ = INPUT$(5, 1)

If five characters were not present in the buffer in the above example, an error would be generated. To avoid this, it is a good idea to ascertain the number of characters in the buffer using LOC (see below) and then to input that number of characters.

The BASIC command GET can also be used with I/O streams. You specify the number of bytes to be read from the buffer in place of the record number that is used when GET is used with files. The number must not be greater than the buffer size specified when the stream was opened.

LOC(file#) returns the number of characters currently in the input buffer for a stream. If the number is greater than 255, BASIC returns 255. To read the characters in the buffer into a variable P$, enter

    P$ = INPUT$(LOC(1), #1)

LOF(file#) returns the amount of free space in the buffer, i.e., the buffer size minus the number of characters currently in the buffer. EOF(file#) indicates whether the buffer is empty, by returning -1 if it is empty and 0 if it is not.

WRITING TO AN I/O STREAM

PRINT #, PRINT # USING, and WRITE # work in communications the same way they work with files. PRINT # sends one variable to the stream. PRINT # USING enables output to be formatted. WRITE # sends a line surrounded by quotation marks. PUT works in a similar way to GET: you specify the number of bytes to be written.

CLOSING AN I/O STREAM

You can close a stream with a CLOSE command, which closes all open files, or a specific CLOSE # command followed by the number given to the stream.

PROGRAMMING THE UART THROUGH BASIC

*D*espite the power of the BASIC communications functions, there will be times when you need direct control. This would include any time you want to send a Break, and any time you want direct control of the handshaking lines.

As we saw in Chapters 13 and 14, the UART contains several registers. Each register contains one byte of information, and each bit within that byte has a different meaning. In order to ascertain the state of the UART, it is necessary to read the registers, and examine the individual bits within them. In order to change the state of the UART, it is necessary to change individual bits within the registers. Therefore, you need to perform the following functions with BASIC:

1. Read a byte from a register

2. Examine the bits within a byte

3. Change the bits within a byte

4. Write a byte out to a register

READING A BYTE FROM A PORT

The BASIC INP command reads a byte from a port. If you refer to Table 14.2, you will see that the modem status register for COM1 is at 3FEH. Therefore, to read the modem status register, enter

MSTATUS = INP (&H3FE)

WRITING A BYTE TO A PORT

The BASIC OUT command sends a byte to a port. The transmitter holding register from COM1 is at 3F8H. To send a character CH to be transmitted, enter

OUT (&H3F8, CH)

BIT MANIPULATION

First, it is necessary to be familiar with Table 15.1, which shows the values of a byte when only one bit is set, for bits zero through seven. I will refer to the numbers beside each bit as the *bit value* of that bit.

Bit	Bit value
0	1
1	2
2	4
3	8
4	16
5	32
6	64
7	128

Table 15.1: Values of bits within a byte

Testing a Bit

You can determine the value of all the bits in a number with the AND command. AND, used with two numbers, produces a new number corresponding to the bits that are set in both numbers. This is known as *Logical AND*. To find out whether a particular bit in a number is set, use AND with the bit value. For example, if STATUS represents the port status and you want to know whether bit 4 of STATUS is set, type

```
IF STATUS AND 16 .....
```

In the above example, if STATUS were 40 decimal, BASIC would make the following computation:

STATUS	0 0 1 0 1 0 0 0
test pattern	0 0 0 1 0 0 0 0
AND result	0 0 0 0 0 0 0 0

This produces a zero result, showing that bit 4 is not set in the status byte. Remember that the rightmost bit is bit zero.

If the STATUS were 52 decimal, the following computation would be made:

STATUS	0 0 1 1 0 1 0 0
test pattern	0 0 0 1 0 0 0 0
AND result	0 0 0 1 0 0 0 0

A nonzero result is produced, showing that bit 4 is set.

Setting a Bit to One

OR, used with two numbers, produces a new number with all bits set that are set in either of the two original numbers. To set a particular bit on within a number, use OR with the bit value. For example, if PARM represents the parameter byte and you want to set bit 3 to one, type

PARM = PARM OR 8

This sets bit 3 on, regardless of its previous state.

If PARM were 37, the computation would be:

PARM	0 0 1 0 0 1 0 1
8	0 0 0 0 1 0 0 0
OR	0 0 1 0 1 1 0 1

Setting a Bit to Zero

NOT, used with a number, produces a new number with each bit the opposite of the corresponding bit in the original number. This is known as the *one's complement*. For example,

NOT 0110

returns

1001

To set a bit within a number to zero, first find the one's complement of the bit value. If the bit is 4, the bit value is 16, which is 00010000. You can find the one's complement of this by typing

NOT 16

which returns

11101111

Next, we AND the original number with this number. This means that every bit in the original number that was one remains one, except for the one corresponding to the zero in the one's complement number.

Thus to turn off bit 5 of the variable PARM, type

```
PARM = PARM AND (NOT 32)
```

If PARM is 106 decimal, we make a mask consisting of NOT 32 as follows:

32	0 0 1 0 0 0 0 0
NOT 32	1 1 0 1 1 1 1 1

Then AND the mask with PARM as follows:

PARM	0 1 1 0 1 0 1 0
NOT 32	1 1 0 1 1 1 1 1
AND	0 1 0 0 1 0 1 0

You will see that 5 has been turned off, which is what we wanted.

EXAMPLES OF BASIC UART CONTROL

Following are a couple of examples of UART control from BASIC. To find out whether the incoming handshaking line DSR is set, enter

```
MSTATUS = INP(&H3FE)
DSR = MSTATUS AND 32
```

To set DTR on and leave the other lines the same, type

```
MCONT = INP(&H3FC)
MCONT = MCONT OR 1
OUT(&H3FC, MCONT)
```

ERROR CONDITIONS

*T*he following are common BASIC error messages relating to communications.

ERROR 69:
COMMUNICATION BUFFER OVERFLOW

Error 69 indicates that characters have been received (i.e., placed into an input buffer by the interrupt system) more quickly than they have been taken out by your program, and that the buffer is filled up. This means that characters have been lost. Possible solutions include trying to use a larger buffer, reducing the baud rate, or implementing flow control such as XON/XOFF or hardware handshaking.

ERROR 25: DEVICE FAULT

Error 25 indicates that an expected handshaking signal was not received. The definition in the BASIC manual is very similar to that for error 24 below, so you should test for both errors.

ERROR 57: DEVICE I/O ERROR

Error 57 is caused by overrun, framing, or parity errors, or the receipt of a break signal. If the word length is set to seven bits, when there is an error the eighth bit of the byte in error is set to one. This indicates where the error occurred.

ERROR 24: DEVICE TIMEOUT

Remember that the OPEN "COM . . ." command can specify timeouts, or periods during which BASIC must wait for certain signals. If periods are not specified, defaults are used. Error 24 indicates that a timeout condition has arisen. For example, suppose your OPEN command had included

CS500

If CTS is not received within 500 milliseconds, error 24 will be generated.

ERROR 68: DEVICE UNAVAILABLE

Error 68 occurs, for example, if you use the command OPEN "COM2..." and you do not have a serial device set up as COM2.

SAMPLE PROGRAM

*T*he program shown in Figure 15.1 is designed to allow you to download files onto an IBM PC from another computer. The other computer must be capable of printing files to a serial printer. It should be set up to print, with a baud rate of 1200, eight data bits, one stop bit, and no parity. If you can choose to transmit a line feed after a carriage return, you should select that option.

I have laid out the program with clarity rather than speed in mind. However, I have tested it at 1200 baud and it works well, whether writing to a floppy disk or to a hard disk. If you need to speed it up, you can try increasing the baud rate, and using a RAM disk to save the file. You can also take out the comments and restructure the program without the GOSUBs. (If you are using COM2 instead of COM1, you should change references to &H3FC to &H2FC.)

In order to create a large enough buffer, you should invoke BASIC by typing

BASICA/S:1024/C:1024

```
10 ' +---------------------------------------------------------+
20 ' |                    CAPTURE.BAS                           |
30 ' |                                                         |
40 ' |           COPYRIGHT 1986 PETER W. GOFTON                |
50 ' |                                                         |
60 ' | INVOKE BASIC AS BASICA/S:1024/C:1024                    |
70 ' |                                                         |
80 ' +---------------------------------------------------------+
100 GOSUB 200: '                        ASK FOR THE FILE NAME
110 GOSUB 300: '                        OPEN THE FILE
120 GOSUB 400: '                        OPEN COMMUNICATIONS
130 GOSUB 500: '                        START MESSAGE
140 GOSUB 600: '                        CAPTURE
150 GOSUB 1000: '                       CLOSE FILES
160 GOSUB 1100: '                       END MESSAGE
170 STOP
200 ' -------------------------------------------------------
210 '              ASK FOR THE FILE NAME
220 '
230 INPUT "ENTER THE NAME OF THE FILE TO CAPTURE: ", F$
240 RETURN
300 ' -------------------------------------------------------
310 '                  OPEN THE FILE
320 '
330 OPEN F$ FOR OUTPUT AS 2
340 RETURN
400 ' -------------------------------------------------------
410 '              OPEN COMMUNICATIONS
420 '
430 OPEN "COM1:1200,N,8,1" AS 1 LEN=1024
440 RETURN
500 ' -------------------------------------------------------
510 '                  START MESSAGE
520 '        .
530 PRINT
540 PRINT "YOU MAY NOW INSTRUCT THE OTHER COMPUTER TO SEND "; F$
550 PRINT "PRESS ESCAPE TO END THE CAPTURE"
560 RETURN
600 ' -------------------------------------------------------
610 '                  CAPTURE
620 '
630 PAUSE = 0: C$ = ""
640 WHILE C$ <> CHR$(27)
650     IF EOF(1) THEN GOTO 710:'        NOTHING RECEIVED
655     B = LOC(1):'                     NUMBER OF CHARS IN BUFFER
660     IF B > 674 THEN GOSUB 830:'      BUFFER > 2/3 FULL: STOP INPUT
670     IF B < 337
        AND PAUSE = 1
        THEN GOSUB 900:'                 BUFFER < 1/3 FULL: RESUME INPUT
680     LN$ = INPUT$(LOC(1), #1):'       READ THE BUFFER INTO CH$
690     PRINT LN$;:'                     DISPLAY IT ON THE SCREEN
700     PRINT #2, LN$;:'                 SAVE IT TO THE FILE
710     C$ = INKEY$:'                    GET ANY KEY HIT
```

Figure 15.1: The CAPTURE.BAS Program

```
720 WEND
730 RETURN
800 ' ------------------------------------------------------
810 '            TURN OFF HANDSHAKING
820 '
830 IF PAUSE = 1 THEN GOTO 870:'      ALREADY IN PAUSE MODE
840 OUT &HF3FC, 0:'                   TURN OFF DTR AND DSR
850 '                                 IN LINE CONTROL REGISTER
860 PAUSE = 1
870 RETURN
900 ' ------------------------------------------------------
910 '            TURN HANDSHAKING BACK ON
920 '
930 OUT &HF3FC, 3:'                   BITS 1 AND 2: DTR AND DSR ON
940 '                                 IN LINE CONTROL REGISTER
950 PAUSE = 0
960 RETURN
1000 ' -----------------------------------------------------
1010 '                    CLOSE
1020 '
1030 CLOSE #1:'                       CLOSE THE COMMS FILE
1040 CLOSE #2:'                       CLOSE THE DISK FILE
1050 RETURN
1100 ' -----------------------------------------------------
1110 '               END MESSAGE
1120 '
1130 PRINT
1140 PRINT "FINISHED CAPTURE"
1150 RETURN
1160 ' -----------------------------------------------------
```

Figure 15.1: The CAPTURE.BAS Program (continued)

Since the IBM PC is normally configured as a DTE device, as are most printers, you will probably be able to use the same cable you use to connect the other computer to a printer: simply replace the other computer's printer with your PC.

You should then run the program. It will ask you for the name of the file you want to download. This must be a valid file name. You will then be told to instruct the other computer to send the file. You should then go to the other computer and issue the appropriate command to print.

The characters received will be echoed to the screen as well as saved to disk. You will see in the screen version that an extra line is left between each line. This is because when printing to the screen a carriage return is treated as a carriage return *and* line feed. The other computer will be sending a carriage return plus line feed at the end of each line. The screen will therefore scroll two lines after each line. However, the disk file will be correct because DOS puts a carriage return and line feed after each line also.

When you see from the screen display that the file has been received, press the ESCAPE key to close the file and exit from the program.

With some slight modifications, this program can make your IBM PC work as a serial to parallel converter. Suppose that you have an IBM PC, a parallel printer, and a CP/M computer with a serial interface but no parallel interface. You want to print out some files from the CP/M computer. You can connect the CP/M computer to the IBM PC, and use the program I have provided; just remove lines 100, 110, and 1040, and change line 700 to

```
700 LPRINT LN$;
```

Each line will then be read from the communications port and printed out.

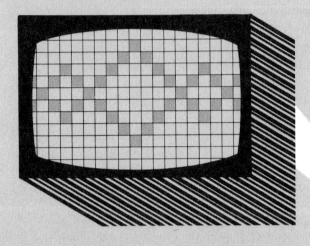

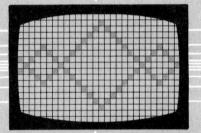

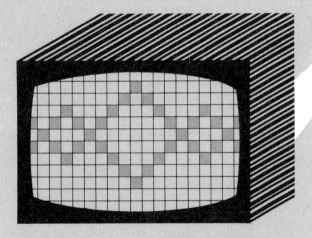

CHAPTER 16

Communicating in C

INTRODUCTION

*P*art Two of this book described the two methods of programming for serial communications on the IBM PC: using operating system calls (DOS and ROM-BIOS) and controlling the hardware at the systems level. This chapter explains how to use the C programming language to program on both these levels, and how to program the UART directly using C. (If you want to write programs in BASIC or assembly language you can go directly to Chapters 15 or 17 respectively. Chapters 15 through 17 are completely self-contained.)

The C language has no standard functions or procedures that are intended for use with serial communications. Accordingly, much of the programming must be done with operating system or hardware level calls. Don't expect much support from the library provided with the compilers, although you can purchase special libraries separately. (They will be discussed later.)

Different compilers have different methods of accessing low-level functions, and accordingly any discussion of C in relation to serial communications must necessarily relate to a specific compiler. This chapter refers to the Microsoft C compiler version 3 for the IBM PC. I will refer to that compiler from now on as *MSC*.

BIT MANIPULATION IN C

*W*hether you use DOS and ROM-BIOS function calls or whether you program the UART directly, you will frequently need to access individual bits within bytes when you are programming for communications, since you will mainly be concerned with reading and setting registers in which different bits have different meanings. Although this is not intended to be a C primer, I have included an overview of how to perform bit manipulation, because that subject is often neglected in standard books on C and is essential to serial communications programming.

First, it is necessary to be familiar with Table 16.1, which shows the values of a byte when only one bit is set, for each bit from zero through seven. I will refer to the numbers beside each bit as the *bit value* of that bit.

Bit	Bit value
0	1
1	2
2	4
3	8
4	16
5	32
6	64
7	128

Table 16.1: Values of bits within a byte

TESTING A BIT

The C operator & is used to perform a logical AND operation. You can determine the value of all the bits in a number with the AND command. AND, used to compare two numbers, produces a new number whose bits correspond to the bits that are set in both numbers.

To find out whether a particular bit is set, use & with the bit value. Thus, if status represents the port status and you want to know whether bit 4 is set, type

```
if (status & 16) {
..........
}
```

In the above example, if status were 40 decimal, C would make the following computation:

status	0 0 1 0 1 0 0 0
test pattern	0 0 0 1 0 0 0 0
AND result	0 0 0 0 0 0 0 0

This produces a zero result, showing that bit 4 is not set in the status byte.

If the status were 52 decimal, C would make the following computation:

status	0 0 1 1 0 1 0 0
test pattern	0 0 0 1 0 0 0 0
AND result	0 0 0 1 0 0 0 0

A nonzero result is produced, showing that bit 4 is set in the status byte.

SETTING A BIT TO ONE

The C operator ¦ (vertical bar) performs a logical OR. OR, used with two numbers produces a new number with all bits set that are set in either of the two original numbers. To set a particular bit on, use ¦ with the bit value. For example, if comparm represents the parameter byte and you want to set bit 3 to one, type

comparm ¦ = 8;

This sets bit 3 on regardless of its previous state.

If comparm were 37, the computation would be:

comparm	0 0 1 0 0 1 0 1
8	0 0 0 0 1 0 0 0
OR	0 0 1 0 1 1 0 1

SETTING A BIT TO ZERO

To set a bit off, we use the one's complement operator, which, in C, is a tilde (˜). This operator sets any bit within a byte that is one to zero, and sets any bit that is zero to one. You perform a logical AND between the original byte and the one's complement of the bit value of the bit to be set off. To turn off bit 5 of the byte parm, for example, you can say

```
parm & =  ~32;
```

If parm is 106 decimal, C would first calculate the byte that is the one's complement of 106 as follows:

32	0 0 1 0 0 0 0 0
~32	1 1 0 1 1 1 1 1

C would then AND this byte with parm as follows:

parm	0 1 1 0 1 0 1 0
~32	1 1 0 1 1 1 1 1
AND	0 1 0 0 1 0 1 0

You can see that bit 5 has now been turned off.

CODING CONVENTIONS FOR BITWISE OPERATORS

By convention, we often use #define statements to indicate the bits within a byte. For an example, refer back to Table 13.6, which gives the meanings of the individual bits within the lines status register of the INS 8250. To use this register in C, you could have something like the following in your header file:

```
#define _DR 1       /* data ready                           */
#define _OE 2       /* overrun error                        */
#define _PE 4       /* parity error                         */
#define _FE 8       /* framing error                        */
#define _BI 16      /* break interrupt                      */
#define _THRE 32    /* transmitter holding register empty   */
#define _TSRE 64    /* transmitter shift register empty     */
```

Assuming that pstatus represents the port status byte you have just retrieved, you can use code such as the following:

```
if (pstatus & _DR)      /* data ready          */
      ......
if (pstatus & _PE)      /* parity error        */
      ......
if (pstatus & _BI)      /* break interrupt     */
      ......
```

OPERATING-SYSTEM CALLS

*M*icrosoft C provides functions with which you can invoke operating system calls. This type of function is sometimes referred to as a *gate*.

THE BDOS( ) FUNCTION

The MSC function bdos() can be used to call most of the standard INT 21H DOS functions. The formal declaration is

```
int bdos(dosfn, dosdx, dosal)
int dosfn;              /* function number to put in AH*/
unsigned int dosdx;     /* value for DX                */
unsigned int dosal;     /* value for AL                */
```

The value the function returns is the contents of the AX register after issuing INT 21H.

In Chapter 12 it was explained that there are two INT21 functions relating to serial communications. Function 3 is for serial input, and function 4 is for serial output. The following C function provides serial input using the DOS INT21 function 3, and returns it as an eight bit unsigned number:

```
unsigned char serin()
    {
```

```
        unsigned char ch;
          ch = (unsigned char) bdos(3, 0, 0);
          return(ch);
        }
```

The following function provides serial output using the DOS INT 21H function 4:

```
        void serout(ch)
        unsigned char ch;
        {
           bdos(4, 0, ch);
        }
```

Notice that no error values are available. The weaknesses of the DOS serial I/O functions are discussed in Chapter 12 and even in the *DOS Technical Reference Manual.*

MSC GATE STRUCTURES

The bdos() function only handles INT 21H. Since the ROM-BIOS serial I/O functions are called through INT 14H, you must use other MSC gate functions. MSC defines three relevant structures that you must understand in order to use these gate functions: WORDREGS, BYTEREGS, and REGS.

1. WORDREGS is a structure that contains values corresponding to the main 8088 registers AX, BX, CX, DX, SI, and DI, and can also hold the carry flag. The elements of the structure, which are all defined as unsigned int, are ax, bx, cx, dx, si, di, and cflag. Each of the four main 16-bit registers AX, BX, CX, and DX can be treated as two separate 8-bit registers. AX can be treated as AH and AL (for high and low), BX can be BH and BL, etc. Loading AL with 34H and AH with 12H is the same as loading AX with 1234H.

2. BYTEREGS is a structure that contains values corresponding to the 8-bit subregisters. They are all defined as unsigned char, and are al, ah, bl, bh, cl, ch, dl, and dh.

3. REGS is a union consisting of x, which is a WORDREGS structure, and h, which is a BYTEREGS structure. To indicate that a value is to be placed in AX, for example, when rg is a union of type REGS, the value can either be loaded as an unsigned integer into rg.x.ax or split into high- and low-order bytes (unsigned chars) and loaded into rg.h.ah and rg.h.al.

THE INT86() FUNCTION

Remember from Chapter 12 that there are four ROM-BIOS functions available for serial communications. These all use interrupt 14H. The function in Program 16.1 uses ROM-BIOS to set the communications parameters. It is assumed that the parameter byte has already been computed and is held in the variable parmbyte.

```
unsigned int setparms(portno, parmbyte)
int portno;          /* 1 for com2, 0 for com1 */
unsigned int parmbyte;
{
union REGS inr, outr;
unsigned int status;

     inr.x.dx = portno;        /* Port Number            */
     inr.h.ah = 0;             /* Function number        */
     inr.h.al = parmbyte;
     int86(0x14, &inr, &outr);  /* ROM-BIOS Serial i/o interrupt */
     status = outr.x.ax;
     return(status);
}
```

Figure 16.1: A function to set parameters using ROM-BIOS

The function in Figure 16.2 calls the ROM-BIOS Transmit Character function and returns the port status byte returned by ROM-BIOS. The function in Figure 16.3 calls the ROM-BIOS Receive Character function and returns the port status byte returned by ROM-BIOS. The function in Figure 16.4 calls the ROM-BIOS Get Port Status function.

```
unsigned char charout(portno, ch)
int portno;                         /* 1 for com2, 0 for com1 */
unsigned char ch;
{
union REGS inr, outr;
unsigned char status;

     inr.x.dx = portno;
     inr.h.ah = 1;                  /* Function number */
     inr.h.al = ch;                 /* character to transmit */
     int86(0x14, &inr, &outr);
     status = outr.h.ah;
     return(status);
}
```

Figure 16.2: A function to transmit characters using ROM-BIOS

```
unsigned char charin(portno, ch)
int portno;                         /* 1 for com2, 0 for com1       */
unsigned char *ch;
{
union REGS inr, outr;
unsigned char status;

     inr.x.dx = portno;
     inr.h.ah = 2;                  /* Function number              */
     int86(0x14, &inr, &outr);
     *ch = outr.h.al;
     status = outr.h.ah;
     return(status);
}
```

Figure 16.3: A function to input characters using ROM-BIOS

```
unsigned int getstat(portno)
int portno;                         /* 1 for com2, 0 for com1       */
{
union REGS inr, outr;
unsigned int status;

     inr.x.dx = portno;
     inr.h.ah = 3;                  /* Function number              */
     int86(0x14, &inr, &outr);
     status = outr.x.ax;
     return(status);
}
```

Figure 16.4: A function to get the port status using ROM-BIOS

SETTING THE PARAMETER BYTE

The function setparms (portno, parmbyte) shown in Figure 16.1 requires the parameter byte to have been computed first. Tables 12.1 through 12.3 gave examples of what bits have to be set to represent different parameters. The function in Figure 16.5 sets any parameters passed to it into a parameter byte. It does not validate the parameters.

```
unsigned char calcparm(baudrate, parity, stop, databits)
int baudrate;          /* 110 through 9600            */
int parity;            /* 0 = None, 1 = Odd, 2 = even */
int stop;              /* 1 or 2                      */
int databits;          /* 7 or 8                      */
{
static int baudlist[] = {110, 150, 300, 600, 1200, 244, 4800, 9600};
register int i = 0;
register int parmbyte;

      for (i = 0; i <= 7; ++i) {        /* For each valid baud rate */
            if (baudlist[i] == baudrate)
                  break;
      }

/* i now equals a number from zero through 7 representing the baud rate */

      parmbyte = i << 5;               /* Shift the value left to the most
                                          significant three bytes      */

      parmbyte |= (parity << 3);       /* Bits 3 and 4 represent
                                          parity                       */

      parmbyte |= ((stop - 1) << 2);   /* Bit 2                        */

      parmbyte |= 2;                   /* Bit 1 always 1               */

      if (databits == 8)
            ++parmbyte;                /* Bit zero                     */

      return(parmbyte);
}
```

Figure 16.5: A function to calculate the parameter byte for ROM-BIOS

PROGRAMMING THE UART DIRECTLY

*I*n order to access the registers of the INS 8250 directly, it is necessary to be able to use the 8088 IN and OUT instructions. MSC provides two functions that do just this:

1. int inp (port) reads one byte from the (unsigned int) port.

2. int outp(port, value) sends the (int) value to the (unsigned int) port and returns value.

Please refer to Table 14.2, which gives the I/O addresses for the various registers for UARTs at COM1 and COM2. You will see that the line status register for COM1 is at 3FDH.

To read it, you can enter

```
unsigned portad = 0x3FD;
int lstatus;
    lstatus = inp(portad);
```

The modem control register for COM2 is at 2FCH. To send a byte modbyte to it, you can enter:

```
unsigned portad = 0x2FC;
    outp(portad, modbyte);
```

Figure 16.6 illustrates setting the parameters (baud rate, etc.) by programming the 8250 directly.

SAMPLE PROGRAMS

*F*ollowing are some complete programs that use MSC to communicate. The user interface has been simplified considerably so as not to complicate matters. Nevertheless, the programs can be compiled and run and may even prove useful to you.

The three programs that follow all use a common startup function fstart(), which appears as Figure 16.7.

```
#define PORTBASE 0x03F8          /* For COM1. 0x02F8 for com2    */

comparm(baudrate, parity, stop, databits)

int baudrate;                    /* 110 through 9600             */
int parity;                      /* 0 = None, 1 = Odd, 2 = even  */
int stop;                        /* 1 or 2                       */
int databits;                    /* 5 through 8                  */
{
unsigned char parmbyte;
unsigned int divisor;
unsigned char lsb, msb;
/* First, calculate the baud rate divisor */

        divisor = 115200 / baudrate;
        msb = divisor >> 8;
        lsb = (divisor << 8) >> 8;

/*
 * Enable access to the divisor latches by setting
 * The divisor latch access bit in the line control register
 */

        outp(PORTBASE + 3, 128);

/* Set the least significant byte */

        outp(PORTBASE, lsb);

/* Set the most significant byte */

        outp(PORTBASE + 1, msb);

/* Start to calculate the parameter byte */

/*
 * First, the word length. This occupies bits 0 and 1 of the parameter byte
 * from 0 0 for 5 bit words to 1 1 for 8 bit words
 */

        parmbyte = databits - 5;

/* Next the stop bits */

        parmbyte |= (stop - 1) << 2;

/* Now the parity */

        if (parity) parmbyte |= 8;            /* parity enable      */

        if (parity == 2) parmbyte |= 16;      /* even               */

        outp(PORTBASE + 3, parmbyte);         /* write it to UART    */
}
```

Figure 16.6: Setting the UART parameters

```
#include "stdio.h";

FILE * fstart(argc, argv)
int argc;
char *argv[];
{
FILE *f;

        if (argc < 2) {
                printf("%s \n", "Must give file name");
                return(NULL);
        }

        f = fopen(argv[1], "w");          /* Open file for writing */

        if (f == NULL) {
                printf("%s \n", "Cannot open file");
                return(NULL);
        }

        return(f);
}
```

Figure 16.7: The FSTART.C function

To use this function, type in Figure 16.7, save it as FSTART.C, and compile it by typing

MSC FSTART;

THE DOWNL1 PROGRAM

The program in Figure 16.8 allows an IBM PC to read files from another computer and save them in a disk file. Only the DOS function to read a character from the AUX port is used. Therefore the program should run on any MS-DOS machine.

To create the program, type it in with the name DOWNL1.C and type

MSC DOWNL1;

Then link it by typing

LINK DOWNL1 FSTART;

```
#include "stdio.h";

main(argc, argv)
int argc;
char *argv[];
{
FILE *f;

        f = fstart(argc, argv);
        if (f == NULL)
                exit();

        printf("%s \n", "Waiting for input");

        downll(f);

        fclose(f);                      /* Close the File        */

        printf("%s \n", "Finished");

}

downll(f)
FILE *f;
{
unsigned char ch;

        for (;;) {
                if (kbhit()) {
                        if (getch() == 27) return;      /* Hit ESC       */
                }

                ch = bdos(3, 0, 0);     /* Get a character       */

                if (ch == 26)           /* ^ Z                   */
                        return;

                fputc(ch, f);           /* record it to the file */

                putchar(ch);            /* Display it            */
        }
}
```

Figure 16.8: DOWNL1.C: Download a file (DOS version)

The program should be able to capture data from any other computer that can direct ASCII output to a serial printer. The other computer will think it is printing. Set up the other computer to print to its serial port, with a low baud rate. Next, use the MODE command to set the parameters of your PC to match the parameters of the other computer (baud rate, stop bits, etc.) as described in Chapter 11.

Run the DOWNL1 program by typing its name followed by the name of the file you want to save. If the file already exists, it will be deleted. For example, if you want to save input in a file called DATA on drive A, type

```
DOWNL1 A:DATA
```

The program will ask you to press any key. Press a key, and then start the other computer "printing." You should see the data displayed on the screen as it comes across.

The problem with the DOS function is that it will wait forever for a character to be received. Accordingly, at the end of the file the program will just "hang up." So I have included a command that tells the program to exit when it receives the character 26, which is Ctrl-Z or end-of-file. You will have to find some way of making the other computer send a Ctrl-Z when the file has ended. You could substitute a different character if you know one that is definitely not used in the file you are transferring. Some computers automatically send Ctrl-Z at the end of a file.

One way of sending Ctrl-Z is to enter BASIC and type

```
LPRINT CHR$(26)
```

Or on a CP/M machine you could enter

```
PIP LST: = EOF:
```

THE DOWNL2 PROGRAM

The program shown in Figure 16.9 is similar to the last program, but uses the ROM-BIOS functions instead of the DOS functions.

It illustrates several common features of serial communications programming, including the concept of a loop in which one or more states are being tested (in this case, whether a key has been

```
#include "dos.h";
#include "stdio.h";

#define PORT 0          /* For COM1. 1 for COM2 */
#define _DATAREADY 1

unsigned char calcparm();
unsigned int setparms();
unsigned char charin();
unsigned int getstat();

main(argc, argv)
int argc;
char *argv[];
{
FILE *f;
unsigned char ch, parmbyte;

        f = fstart(argc, argv);
        if (f == NULL)
                exit();

/* Calculate and set the parameter byte */

        parmbyte = calcparm(300, 0, 1, 8);    /* Figure 16.5      */
        setparms(PORT, parmbyte);             /* Figure 16.1      */

/*
* First we have to call charin because only that sets up the outgoing
* handshaking lines. It will return after a timeout.
*/
        charin(PORT, &ch);

        printf("%s \n", "Press any key to start download");
        getch();

        downl2(f);

        fclose(f);                            /* close the file   */

        printf("%s \n", "Finished");

}

downl2(f)
FILE *f;
{
unsigned char ch, ah, al;
unsigned int status, timeout;

        printf("%s \n", "Press 'ESC' to stop download");

        for (;;) {
```

Figure 16.9: DOWNL2.C: Download a file (ROM-BIOS version)

```
                    if (kbhit()) {
                            if (getch() == 27)
                                    return;            /* Hit ESC            */
                    }

                    if (timeout == 4000) {             /* Warning message    */
                            printf("%s \n", "Waiting");
                            timeout = 0;
                    }

                    status = getstat(PORT);            /* Figure 16.4        */

                    ah = status >> 8;

                    al = status;
    /*
    * We could test the other status bits for errors here
    */
                    if (! (ah & _DATAREADY)) {         /* Nothing received   */
                            ++timeout;
                            continue;
                    }

                    timeout = 0;
    /* We know a character has been received so we will read it               */

                    charin(PORT, &ch);                 /* Figure 16.3        */

                    fputc(ch, f);                      /* record it to the file */

                    putchar(ch);                       /* display it         */
            }
    }
```

Figure 16.9: DOWNL2.C: Download a file (ROM-BIOS version) (continued)

hit and whether a character has been received) and processing accordingly. In other words, it illustrates polling. The program also shows how you can set the parameters from within a program using the ROM-BIOS functions, and do not have to rely on the user running the MODE program correctly.

You will also see how much more convenient it is to terminate the program at the receiving end by pressing a key rather than having to rely on receiving a specific end-of-file character from the remote device. You cannot do this with DOS because you can't poll the input port and the keyboard in turn.

THE DOWNL3 PROGRAM

DOWNL3, shown in Figure 16.10, again has the same purpose as the last two programs, but this time uses the direct control method of IN and OUT instructions.

```c
#include "stdio.h";

#define _DATAREADY 1
#define _OVERRUN 2
#define _PARITY 4
#define _FRAME 8
#define _BREAK 16
#define _DTR 1
#define _RTS 2
#define PORTBASE 0x02F8 /* For COM1. 0x03F8 for COM2    */
#define BUFSIZE 512
main(argc, argv)
int argc;
char *argv[];
{
FILE *f;
unsigned char parmbyte;
int count;
char buf[BUFSIZE];

        f = fstart(argc, argv);
        if (f == NULL)
                exit();
/* Calculate and set the parameter byte */

        comparm(1200, 0, 1, 8);                 /* Figure 16.6  */

        printf("%s \n", "Press any key to start download");
        getch();

        printf("%s \n", "Press 'ESC' to stop download");

        while (count = readbuf(buf) == BUFSIZE) {
                if (count > 0)
                        fwrite(buf, count, 1, f);
        }

        fclose(f);                              /* close the file        */

        printf("%s \n", "Finished");
readbuf(buf)
char *buf;
{
```

Figure 16.10: Download a file by programming the UART

```
int count = 0;
int lstat;                                        /* line status        */
int mstat;                                        /* modem status       */

        outp(PORTBASE + 4, _DTR | _RTS);          /* handshaking on     */

        while (count <= BUFSIZE) {

                if (kbhit()) {
                        if (getch() == 27)
                                return;           /* hit ESC            */
                }

                lstat = inp(PORTBASE + 5);        /* read line status   */

                if (lstat & _OVERRUN) {
                        printf("%s \n", "overrun error");
                        return(-1);
                }

                if (lstat & _PARITY) {
                        printf("%s \n", "parity error");
                        return(-1);
                }

                if (lstat & _FRAME) {
                        printf("%s \n", "framing error");
                        return(-1);
                }

                if (lstat & _BREAK) {
                        printf("%s \n", "break received");
                        return(-1);
                }
/*
 * Read the modem status (not needed but here for completeness)
 * We can test for incoming handshaking if we wish.
 */
                mstat = inp(PORTBASE + 6);

                if (!(lstat & _DATAREADY))         /* nothing there      */
                        continue;
/* So we have received a character */

                buf[count++] = inp(PORTBASE);      /* read the character */
        }

/* We have filled the buffer, or hit ESCAPE */

        outp(PORTBASE + 4, 0);                     /* handshaking off    */

        return(count);
}
```

Figure 16.10: Download a file by programming the UART (continued)

You will recall that under DOS and BIOS you have no control over the handshaking signals, and so you can't prevent the remote device from sending the whole file at once. This limits the baud rate, since the program saves the characters to a file, and the process has to move slowly so as not to lose any incoming characters while saving to disk.

With DOWNL3, we turn on the handshaking signals, receive 512 bytes into memory, turn off the handshaking signals, save the 512 bytes, and so on. Since the program is no longer receiving and saving at the same time, it can operate at a faster baud rate.

INTERRUPT DRIVEN I/O

*I*n order for characters to be received and transmitted upon the occurrence of interrupts generated by the UART, it is necessary to construct an interrupt handler as mentioned in Chapter 14.

Because of the way that the interrupt handler must work, you cannot write it entirely in C. Part of it must be written in assembly language—a task for experienced assembly language programmers. Chapter 18 presents some guidelines for doing so and includes an assembly language function that intercepts serial I/O interrupts and automatically calls a C function written by you when they occur.

If you do not want to write your own routines, I strongly recommend the software package Asynch Manager from Blaise Computing that consists of a library of functions that you can link in with your software. These functions set up interrupt driven I/O, automatically load a buffer with incoming characters, and read a buffer for characters to be sent. Other functions are available, including sending and receiving strings of characters and the use of the XMODEM protocol. I have used the package myself for two different products, and would much rather buy an existing package than rewrite the functions. There are other packages, which may be equally good, but the Blaise set is the one with which I am familiar (they also publish a set for Turbo Pascal).

Asynch Manager is available from:

Blaise Computing Inc.
2034 Blake Street
Berkeley, CA 94704
(415) 540-5441

CIRCULAR BUFFERS

*T*he concept of circular buffers was described in Chapter 10. An area of memory is set up into which characters are placed and from which characters are removed. Typically, an interrupt mechanism places received characters into the buffer, and an application program takes them out. The program in Figure 16.11 illustrates how a circular buffer can be created in C.

```
typedef     struct {
    int count;          /* number of chars now in the buffer     */
    int start;          /* offset of next characte to take        */
    int size;           /* size of the buffer                     */
    char *buffer;       /* address of the buffer                  */
} RING;

/*                                                                  */
void initbuf(r, addr, len)
register RING *r;
char *addr;
int len;

/*
 * initialize a RING buffer
 */

{
    r -> count   = 0;
    r -> start   = 0;
    r -> buffer  = addr;
    r -> size    = len;
}
/*                                                                  */
void putbuf(r, ch)
register RING *r;
char    ch;
```

Figure 16.11: Circular buffer creation

```
/*
 * put a character into a RING buffer
 */

{
int offset;

    offset = (r -> start + r -> count) % r -> size;
                                                    /* position for new char  */

    r -> buffer[offset] = ch;                       /* place char in buffer   */

    if (r -> count >= r -> size) {
                                                    /* overflow?              */
        r -> start++;                               /* move starting point    */

        if (r -> start >= r -> size)                /* start is beyond end    */
            r -> start -= r -> size;                /* rotate it              */

    }

    else {
        r -> count++;                               /* just update the count  */
    }
}
/*_____ */
int getbuf(r, ch)
register RING      *r;
char *ch;
/*
 * get a character from a RING buffer
 * returns 0 if buffer is empty.
 */
{

    if (r -> count == 0)                            /* nothing there          */
        return (0);                                 /* return 'failure'       */

    *ch = r -> buffer[r -> start];                  /* get next char to take  */

    r -> start++;                                   /* move starting point    */

    if (r -> start >= r -> size) {                  /* do not overflow        */
        r -> start = 0;
    }

    r -> count--;                                   /* one less char there    */

    return(1);                                      /* return 'success'       */
}
/*_____ */
```

Figure 16.11: Circular buffer creation (continued)

CRC COMPUTATIONS

*T*he final function presented here calculates the CRC-CCITT used in some versions of XMODEM, in YMODEM, and in Kermit. (See Chapter 10 for an explanation of the principle of CRC calculations.) The function in Figure 16.12 comes from a public domain program, and I am grateful to Chuck Forsberg and Stephen Satchell for drawing it to my attention.

The variable crc should be initialized to zero, and the function should be called for each character in the string.

```
unsigned onecrc(crc, cp)
register unsigned int crc;
char cp;

{
register int count;
unsigned c;

    c = cp & 255;

    for (count = 8; --count >= 0; ) {
        if (crc & 0x8000) {
            crc <<= 1;
            crc += (((c <<= 1) & 0400) != 0);
            crc ^= 0x1021;
        }

        else {
            crc <<= 1;
            crc += (((c <<= 1) & 0400) != 0);
        }
    }

    return(crc);
}
```

Figure 16.12: CRC-CCITT calculation function

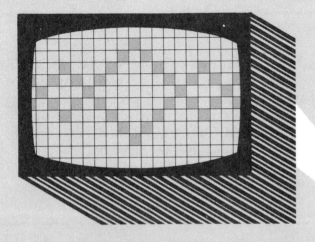

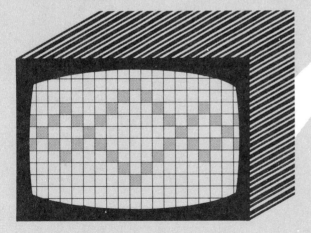

CHAPTER 17

Communicating in Assembly Language

INTRODUCTION

*T*his chapter is intended for advanced programmers who want to use assembly language for serial communications programming. It begins with a description of bit manipulation, and then discusses the use of DOS and ROM-BIOS function calls, and ways of programming the hardware directly.

BIT MANIPULATION IN ASSEMBLY LANGUAGE

*W*hether you use DOS and ROM-BIOS function calls or whether you program the hardware directly, you will frequently need to access individual bits within bytes, since a lot of information is stored in this way by the UART. First, you should be familiar with Table 17.1, which shows the values of a byte when only one bit is set, for each bit from zero through seven. I will refer to the numbers beside each bit as the *bit value* of that bit.

TESTING A BIT

In assembly language, you can use the TEST instruction to find out whether a particular bit in a value is set. TEST performs

Bit	Bit value
0	1
1	2
2	4
3	8
4	16
5	32
6	64
7	128

Table 17.1: The values of bits within a byte

a logical AND internally, but does not change the contents of the registers. It sets the zero flag if a zero result is produced. You TEST the value to be examined with the bit value of the bit to be examined. For example, suppose that a byte represented by STATUS represents the port status and you want to know whether bit 4 is set. Type

```
MOV AL, STATUS
TEST AL, 10H
JZ NOTSET
; if we get here, the bit is set

........

........
NOTSET:
; if we jump to here, the bit is not set
```

In the above example, if STATUS were 40H, the following computation would be made:

STATUS	0 0 1 0 1 0 0 0
test pattern	0 0 0 1 0 0 0 0
AND result	0 0 0 0 0 0 0 0

This produces a zero result (the zero flag is set), showing that bit 4 is not set in the status byte.

If the STATUS were 52 decimal, the following computation would be made:

STATUS	0 0 1 1 0 1 0 0
test pattern	0 0 0 1 0 0 0 0
AND result	0 0 0 1 0 0 0 0

A nonzero result is produced, (the zero flag is not set) showing that bit 4 of STATUS is set.

SETTING A BIT TO ONE

The operator OR performs a logical OR. One number ORed with another produces a new number with all bits set that are set in either of the original two numbers. To set a particular bit on, use OR with the bit value. For example, if COMPARM represents the parameter byte and you want to set bit 3 to one, you type

```
MOV AL, COMPARM
OR  AL, 08H
```

This sets the bit on regardless of its previous state.

If COMPARM were 37 decimal, the computation would be

COMPARM	0 0 1 0 0 1 0 1
8	0 0 0 0 1 0 0 0
OR	0 0 1 0 1 1 0 1

SETTING A BIT TO ZERO

To set a bit to zero, use the NOT operator to find the one's complement of the bit value, and AND the result with the original byte. The NOT operator sets any bit within a byte that is one to zero, and sets any bit that is zero to one.

To turn off bit 5 of the byte PARM you could type

```
MOV AL, 20H
NOT AL
AND PARM, AL
```

If PARM were 106 decimal, the computer would calculate the one's complement of PARM as follows:

32	0 0 1 0 0 0 0 0
NOT 32	1 1 0 1 1 1 1 1

It would then AND this byte with PARM as follows:

PARM	0 1 1 0 1 0 1 0
NOT 32	1 1 0 1 1 1 1 1
AND	0 1 0 0 1 0 1 0

You can see that as a result bit 5 has been turned off, as you wanted. For simplicity, with most assemblers you can enter

```
MOVE AL, NOT 20H
AND PARM, AL
```

The assembler itself would compute NOT 20H for you.

USING DOS INTERRUPTS FROM ASSEMBLY LANGUAGE

*R*emember from Part Two that there are two DOS interrupt functions that relate to serial communications. The first function sends a character to the standard auxiliary device. It is achieved as follows:

```
MOV DL, CH_TO_SEND
MOV AH, 04H  ; DOS function call number
INT 21H        ; standard DOS interrupt
```

The second function receives a character from the standard auxiliary devices. It is achieved as follows:

```
MOV AH, 03H  ; DOS function call number
INT 21H        ; standard DOS interrupt
MOV CH_RECEIVED, AL
```

USING ROM-BIOS INTERRUPTS FROM ASSEMBLY LANGUAGE

*T*he first ROM-BIOS call sets the communications parameters. The parameters must have been stored in a parameter byte, which we will call PARMBYTE. We assume that STATUS has been defined as a two-byte area of memory. The sequence is

```
MOV DX, PORT_NO      ; 0 for COM1, 1 for COM2
SUB AH, AH           ; set AH to zero
```

```
MOV AL, PARMBYTE      ; a byte containing the parameters
INT 14H               ; ROM-BIOS serial I/O interrupt
MOV STATUS, AX        ; status information
```

The next call sends a character to the serial port:

```
MOV DX, PORT_NO       ; 0 for COM1, 1 for COM2
MOV AH, 1             ; function to send a character
MOV AL, CH_TO_SEND    ; character to send
INT 14H               ; ROM-BIOS serial I/O interrupt
MOV LSTATUS, AH       ; line status
```

The next call receives a character from the serial port:

```
        MOV DX, PORT_NO       ; 0 for COM1, 1 for COM2
        MOV AH, 2             ; function to receive a character
        INT 14H               ; ROM-BIOS serial I/O interrupt
        TEST AH, AH           ; is AH zero?
        JNZ ERROR             ; if not, we have an error
        MOV CH_RECD, AL       ; character received
        RET
ERROR:
        MOV CH_RECD, 0
        RET
```

The calling function should examine AH to see whether bit seven is on (indicating timeout) or whether other bits are on (indicating line status errors).

The final call gets the line status byte and the modem status byte. I will assume that you have declared LSTATUS and MSTATUS appropriately.

```
MOV DX, PORT_NO       ; 0 for COM1, 1 for COM2
MOV AH, 3             ; function to get status
INT 14H               ; ROM-BIOS serial I/O interrupt
MOV LSTATUS, AH       ; line status byte
MOV MSTATUS, AL       ; modem status byte
```

CONTROLLING THE UART IN ASSEMBLY LANGUAGE

*T*he addresses of the UART registers were given in Chapters 13 and 14. To read and write values, the OUT and IN commands are used. The command to send a byte to a port is the OUT command. It is normally coded as

```
MOV DX, PORT_ADR      ; address of the port
MOV AL, TO_SEND       ; byte to send to the port
OUT DX, AL            ; send the byte to the port
```

The command to receive a byte from a port is the IN command. It is normally coded as

```
MOV DX, PORT_ADR      ; address of the port
IN AL, DX             ; AL now contains the byte read
```

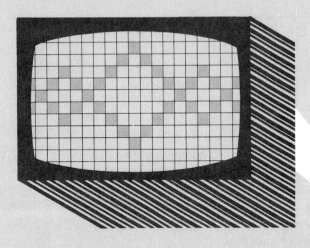

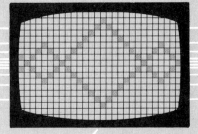

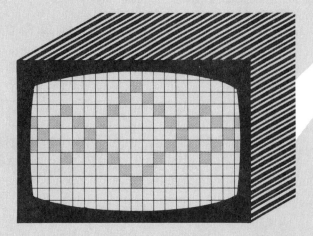

CHAPTER 18

An Interrupt Service Routine

INTRODUCTION

Chapter 14 covered the operation of interrupt-driven communications. This chapter contains a specific example of an interrupt service routine (ISR) written in a combination of C and assembly language. An ISR is a section of code that is called automatically when an interrupt occurs. This chapter is intended for advanced programmers only.

OVERVIEW

There are three functions provided here. Two of them are written in assembly language and one is written in C.

1. *intinit* is an assembly language function that initializes the system for interrupt-driven communications.

2. *inthand* is an assembly language function that is called automatically whenever a comunications interrupt occurs, and is a gate to cfunct().

3. *cfunct()* is a C function that deals with the interrupt.

The functions have been tested with the Microsoft Macro Assembler version 4.0 and the Microsoft C compiler version 3.0 (small memory model).

INITIALIZATION

Figure 18.1 contains the intinit function, which is called at the beginning of the program that is running in conjunction with the ISR. This function initializes both the UART and the PIC (programmable interrupt controller), and sets up a pointer in the PC's interrupt vector table to inthand.

```
; INTINIT.ASM

; Copyright (c) 1986 Peter W. Gofton
; Interrupt handler for IBM PC serial I/O
; Designed to interface with Microsoft C
; Assemble as: MASM INTINIT;

; This function is called from C as intinit().
; It sets up an interrupt handler to be called automatically when
; an I/O interrupt occurs.


;---------------------- EQUATES ------------------------------------

PORT_AD    equ 03F8H           ; 02F8H for COM2
RX         equ 03F8H           ; 02F8H for COM2
TX         equ 03F8H           ; 02F8H for COM2
INT_EN     equ 03F9H           ; 02F9H for COM2
INT_ID     equ 03FAH           ; 02FAH for COM2
LCONT      equ 03FBH           ; 02FBH for COM2
MCONT      equ 03FCH           ; 02FCH for COM2
LSTAT      equ 03FDH           ; 02FDH for COM2
MSTAT      equ 03FEH           ; 02FEH for COM2
COM_INT    equ 0CH             ; 0BH for COM2


;-------------------- INITIAL DECLARATIONS -------------------------

_TEXT    SEGMENT   BYTE PUBLIC 'CODE'
_TEXT    ENDS

CONST    SEGMENT   WORD PUBLIC 'CONST'
CONST    ENDS

_BSS     SEGMENT   WORD PUBLIC 'BSS'
_BSS     ENDS

_DATA    SEGMENT   WORD PUBLIC 'DATA'
_DATA    ENDS

DGROUP   GROUP    CONST,   BSS,    DATA
         ASSUME  CS: _TEXT, DS: DGROUP, SS: DGROUP, ES: DGROUP


_TEXT SEGMENT

EXTRN _inthand: NEAR

PUBLIC _myds
_myds DW "DS"

PUBLIC _intinit
```

Figure 18.1: The intinit function

```
_intinit      PROC NEAR

;--------------------------------------------------------------------


;------------------------ SAVE DS in a CS location ------------------------

      PUSH    SI
      MOV     SI, OFFSET _myds
      MOV     CS:[SI], DS
      POP     SI

;--------------------- SET THE INTERRUPT VECTOR ---------------------------


      PUSH    DS            ; save DS

      PUSH    CS            ; to load into CS
      POP     DS            ; load CS into DS

      LEA     DX, _inthand  ; address of the interrupt handler
      MOV     AH, 025H      ; function call to set vector
      MOV     AL, COM_INT   ; vector to set
      INT     21H           ; DOS interrupt

      POP     DS            ; restore DS


;------------------- ENABLE INTERRUPTS ON THE PIC ------------------------

      CLI

      MOV     DX, PORT_AD
      ADD     DX, 4         ; modem control register
      MOV     AL, 0FH       ; DTR, RTS, OUT1, OUT2
      OUT     DX, AL        ; enable handshaking

      MOV     DX, 21H       ; address of the PIC
      IN      AL, DX        ; get contents
      AND     AL, NOT 10H   ; set off the bit for IRQ4. NOT 8 for IRQ3
      OUT     DX, AL        ; send it out to the controller

      STI

;--------------------- RESET THE SERIAL INTERFACE ------------------------


      MOV     DX, INT_EN
      MOV     AL, 0
      OUT     DX, AL
```

Figure 18.1: The intinit function (continued)

```
        MOV     DX, LSTAT
        IN      AL, DX

        MOV     DX, MSTAT
        IN      AL, DX

        MOV     DX, RX
        IN      AL, DX

        MOV     DX, INT_ID
        IN      AL, DX


;-------------------- ENABLE INTERRUPTS ON THE UART ---------------------

; Now we enable the interrupts on the serial interface

        MOV     DX, INT_EN   ; address of interrupt enable register
        MOV     AL, 0FH      ; code to enable all four interrupts
        OUT     DX, AL       ; set it

;---------------------- RESET THE PIC -----------------------------------

        MOV     DX, 20H      ; address of the PIC
        MOV     AL, 20H      ; reset code
        OUT     DX, AL       ; set it

;---------------------- END THE FUNCTION --------------------------------

        RET

_intinit endp

_TEXT ends

END

;-----------------------------------------------------------------------
```

Figure 18.1: The intinit function (continued)

Under the small memory model, your C program accesses its data by referencing a single data segment. This is stored in the DS register of the CPU. However, from time to time DS can be changed temporarily while a DOS function or another ISR is executing. For this reason, your ISR may be called at a time when

DS does not contain the correct data segment for your C program. Accordingly, if inthand merely passed control to cfunct(), cfunct() might not be able to find its data.

The inthand function cannot even find out what that data segment is, because only the CS and IP registers are restored by the PC's interrupt mechanism. For this reason, intinit saves C's data segment in a variable within the code segment. This allows inthand to retrieve the data segment for cfunct(), and restore it before calling cfunct().

Similarly, inthand could not locate C's stack segment. Instead, intinit also allocates a special temporary stack segment to be used by inthand.

PART ONE OF THE ISR: INTHAND

*T*he inthand function shown in Figure 18.2 is called whenever a communications event occurs. It serves as a gate to cfunct(), which is a function written in C.

```
;  INTHAND.ASM

;  Copyright (c) 1986 Peter W. Gofton

;  Interrupt handler for IBM PC Serial I/O
;  Designed to interface with Microsoft C
;  Assemble as: MASM INTHAND;
;  Read in conjunction with the text


;------------------- INITIAL DECLARATIONS -------------------

_TEXT    SEGMENT   BYTE PUBLIC 'CODE'
_TEXT    ENDS

ASSUME   CS: _TEXT, DS: NOTHING, SS:NOTHING, ES:NOTHING


_TEXT        SEGMENT
```

Figure 18.2: The inthand function

```
; Create a stack for use by the C function

STACKSIZE equ 256
intstack DB STACKSIZE DUP (?)

; External declarations

EXTRN _myds:WORD

EXTRN _cfunct:NEAR

public _tempsp
_tempsp dw "SP"

public _tempss
_tempss dw "SS"


PUBLIC _inthand

_inthand    PROC FAR


;------------------- PREPARE TO CALL THE C FUNCTION ---------------------

    STI                     ; enable interrupts

    PUSH    AX              ; save all registers
    PUSH    BX
    PUSH    CX
    PUSH    DX
    PUSH    SI
    PUSH    DI
    PUSH    DS
    PUSH    ES

    MOV     AL, 20H
    OUT     20H, AL         ; reset the programmable interrupt controller

    MOV     SI, OFFSET _myds
    MOV     DS, CS:[SI]   ; restore data segment for the C code

    MOV     _tempss, SS   ; save stack segment
    MOV     _tempsp, SP   ; and stack pointer

    PUSH    CS
    POP     SS              ; use code segment for stack

    MOV     SP, offset intstack + STACKSIZE - 1


; ------------------- CALL THE C FUNCTION ---------------------------
```

Figure 18.2: The inthand function (continued)

```
; We are now ready to call the C function we wrote to handle the
; communications events.

    CALL     _cfunct      ; Call the C function

; -------------------- RESTORE THE STATUS -----------------------------

; We have now returned from our C function and must restore the
; registers to their original values.

    MOV      SP, _tempsp  ; restore the stack pointer
    MOV      SS, _tempss  ; and stack segment

    POP      ES           ; restore all registers
    POP      DS
    POP      DI
    POP      SI
    POP      DX
    POP      CX
    POP      BX
    POP      AX

    IRET

_inthand endp
;---------------------------------------------------------------------
_TEXT     ENDS

END
;---------------------------------------------------------------------
```

Figure 18.2: The inthand function (continued)

The inthand function starts by issuing an STI command to set the interrupt flag in the CPU so that another interrupt can occur. It then saves the current registers on the stack. Next it resets the PIC, so that the PIC can generate further interrupts. This is done by sending 20H to port 20H.

The inthand function then restores the DS for the C part of the program, saves the current SS (stack segment) and SP (stack pointer), and sets up a temporary stack segment and stack pointer for use by cfunct(). It then calls cfunct(), restores the environment existing at the time it was called, and exits with an IRET (interrupt return) command.

PART TWO OF THE ISR: CFUNCT( )

*F*igure 18.3 contains cfunct(), which is a continuation of the ISR. It is written in C so as to allow easier modification and access to external variables declared in your C code.

```
/*-------------------------------- CFUNCT.C ----------------------------*/
int mstat;                              /* modem status                 */
int lstat;                              /* line status                  */
int comerror;                           /* error number                 */

#define PARITY 1
#define FRAME 2
#define BREAK 3

#define COMBASE 0x03F8                  /* 0x02F8 for COM2              */

#define TXREG   COMBASE                 /* transmit reg                 */
#define RXREG   COMBASE                 /* receive reg                  */
#define INTID   COMBASE + 2             /* interrupt ID reg             */
#define LINSTAT COMBASE + 5             /* line Status reg              */
#define MODSTAT COMBASE + 6             /* modem Status reg             */

typedef struct {
    int count;                          /* number of chars now in the buffer*/
    int start;                          /* offset of next character to take */
    int size;                           /* size of the buffer           */
    char *buffer;                       /* address of the buffer        */
} RING;

RING ibuf;                              /* input buffer                 */
RING obuf;                              /* output buffer                */

extern void putbuf();                   /* see Figure 16.11             */
extern int getbuf();                    /* see Figure 16.11             */

/*--------------------------------------------------------------------*/
void cfunct()
{
int intnr, ch;

    for (;;) {                          /* until no more ints           */

        intnr = inp(INTID);             /* interrupt ID                 */

        if (intnr & 1) {                /* no interrupt pending         */
            return;
        }
```

Figure 18.3: The cfunct() function

```
    switch (intnr) {

        case(0):                    /* modem status                  */
            mstat = inp(MODSTAT);   /* read the status               */
            break;

        case(2):                    /* ready to transmit             */
            if (getbuf(&obuf, &ch)) /* if we have one to send        */
                outp(TXREG, ch);    /* send it                       */
            break;

        case(4):                    /* received data                 */
            ch = inp(RXREG);        /* read the char                 */
            putbuf(&ibuf, ch);      /* save it in buffer             */
            break;

        case(6):                    /* line status                   */
            lstat = inp(LINSTAT);
            if (lstat & 4)          /* parity error                  */
                comerror = PARITY;
            else if (lstat & 8)     /* framing error                 */
                comerror = FRAME;
            else if (lstat & 16)    /* break interrupt               */
                comerror = BREAK;
            break;
        }
    }
}
/*-----------------------------------------------------------------*/
```

Figure 18.3: The cfunct() function (continued)

The cfunct() function first examines the interrupt identification register of the UART in order to see what event triggered the interrupt. It then acts appropriately depending on the cause of the interrupt. If the interrupt was caused by the receipt of a character, the character is read and saved in a circular buffer. If it was caused by an error, an error flag is set. If it was caused by successful transmission of a character, the next character (if any) is sent.

COMPILING CFUNCT()

The cfunct() function should be compiled using the /Gs option in order to suppress stack checking. It is impossible to use your program's own stack for the interrupt handler, and inthand allocates a new one. For this reason, C's stack overflow checking mechanism reports an error at run-time and aborts the program unless stack checking is suppressed.

The normal way to compile cfunct() is as follows:

```
MSC /Gs cfunct;
```

The cfunct() function calls the circular buffer functions provided in Chapter 16. These should also be compiled with the /Gs option because the temporary stack will still be in effect when they are called.

MODIFYING CFUNCT( )

You may well want to change cfunct() to suit your particular requirements. When doing so, keep the following points in mind:

1. Because DOS is not re-entrant, cfunct() cannot call any DOS functions or any functions that call DOS functions. If your ISR is triggered while a DOS function is in operation, and your ISR attempts to call another DOS function, DOS will probably crash the system.

2. Make sure that cfunct() does not allocate too many local variables because it could overflow the stack we have provided. You can provide for a larger stack by changing the STACKSIZE equate in intinit.

PUTTING IT TOGETHER

*A*s an example of how these functions work together, look at Figure 18.4. This program creates two circular buffers (using the functions provided in Chapter 16), calls intinit to set up the communications interrupts, and then waits for characters to be available in the input buffer, printing them as they appear. Incidentally, there is no problem linking INTHAND.OBJ and INTINIT.OBJ with a C program. The names can simply be added to the list of object modules passed to the linker.

```c
/*----------------------- INTTEST.C ----------------------------------*/
#include "stdio.h";

typedef struct {
    int count;                      /* number of chars now in the buffer*/
    int start;                      /* offset of next character to take */
    int size;                       /* size of the buffer               */
    char *buffer;                   /* address of the buffer            */
} RING;

extern RING ibuf;                   /* input buffer                     */
extern RING obuf;                   /* output buffer                    */

char *calloc();
/*-------------------------------------------------------------------*/
main()
{
int chgot;
char *ispace, *ospace;

    ispace = calloc(256, 1);        /* space for input buffer           */
    ospace = calloc(256, 1);

    initbuf(&ibuf, ispace, 256);    /* see Figure 16.11                 */
    initbuf(&obuf, ospace, 256);

    intinit();                      /* intinit.asm                      */

    comparm(1200, 0, 1, 8);         /* see Figure 16.5                  */

    printf("%s \n", "Done intinit");

    for (;;) {

        if (kbhit()) {              /* user can terminate with esc      */
            if (getch() == 27)
                exit();
        }

        if (getbuf(&ibuf, &chgot)) { /* we have got a character          */
            fputc(chgot, stdout);    /* show it                          */
        }
    }
}
/*-------------------------------------------------------------------*/
```

Figure 18.4: Test Program for Interrupt Service Routine

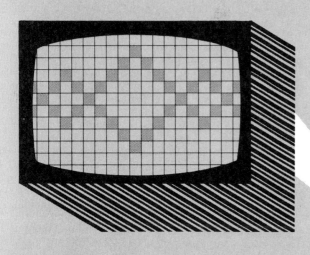

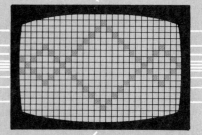

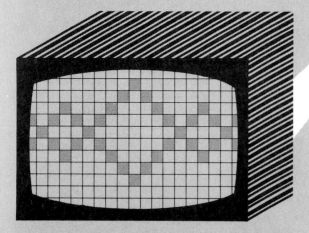

APPENDIX *A*

Common Pin
Connections

The full set of RS-232 circuits

Pin	Circuit	Abbreviation	Full name	Direction
Data:				
2	BA	TXD	Transmitted data	DTE to DCE
3	BB	RXD	Received data	DCE to DTE
Primary handshaking lines:				
6	CC	DSR	Data set ready	DCE to DTE
20	CD	DTR	Data terminal ready	DTE to DCE
Secondary handshaking lines:				
4	CA	RQS	Request to send	DTE to DCE
5	CB	CTS	Clear to send	DCE to DTE
Modem lines:				
8	CF	CD	Carrier detect	DCE to DTE
22	CE	RI	Ring indicator	DCE to DTE
Ground or common:				
7	AB	SG	Signal ground	
Less commonly used circuits:				
1	AA		Protective ground	
12	SCF		Secondary received line signal det.	DCE to DTE
13	SCB		Secondary clear to send	DTE to DCE
14	SBA		Secondary Transmitted Data	DTE to DCE
15	DB		Transmitter signal element timing	DCE to DTE
16	SBB		Secondary received data	DCE to DTE
17	DD		Receiver signal element timing	DCE to DTE
19	SCA		Secondary request to send	DTE to DCE
21	CG		Signal quality detector	DCE to DTE
23	CH		Data signal rate selector	DTE to DCE
23	CI		Data signal rate selector	DCE to DTE
24	DA		Transmitter signal element timing	DTE to DCE

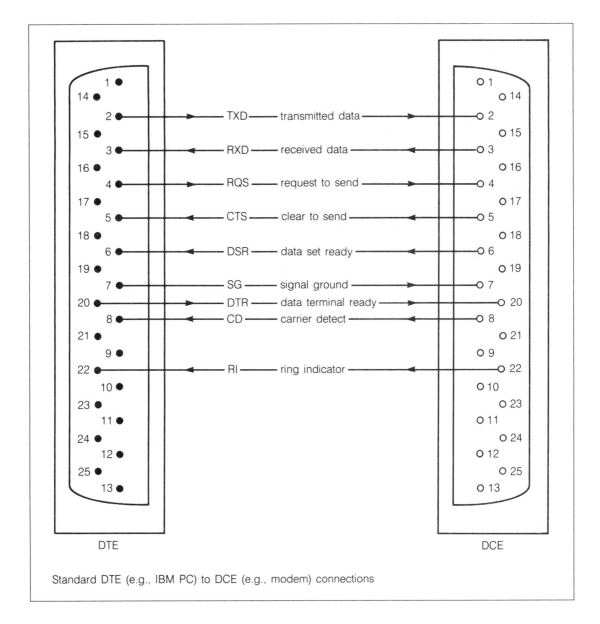

DTE

DCE

Standard DTE (e.g., IBM PC) to DCE (e.g., modem) connections

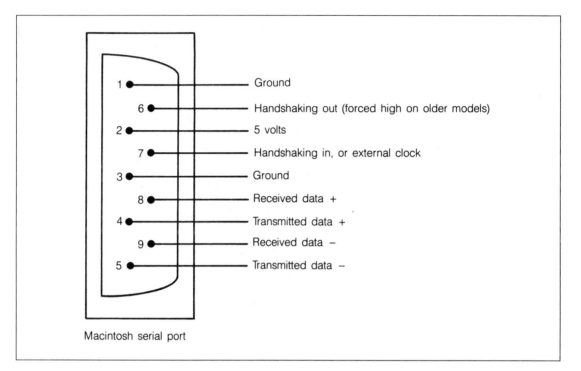

Macintosh serial port

Pin	Signal
1	Ground
6	Handshaking out (forced high on older models)
2	5 volts
7	Handshaking in, or external clock
3	Ground
8	Received data +
4	Transmitted data +
9	Received data −
5	Transmitted data −

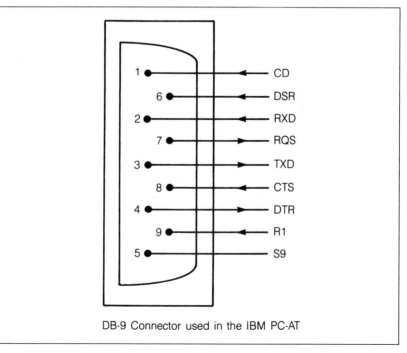

Pin	Signal
1	CD
6	DSR
2	RXD
7	RQS
3	TXD
8	CTS
4	DTR
9	R1
5	S9

DB-9 Connector used in the IBM PC-AT

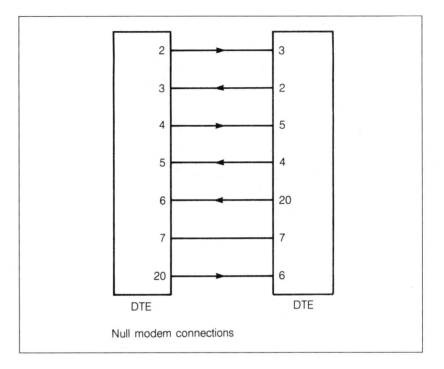

Null modem connections

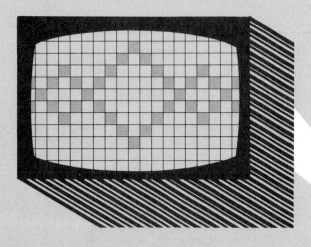

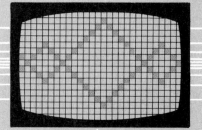

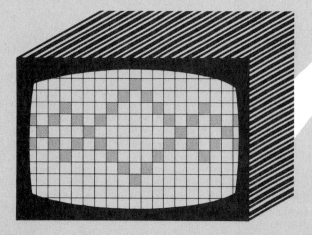

Glossary

ACK   Acknowledge, ASCII code 6. Used to acknowledge receipt of a transmission.

Acoustic coupler   A modem connected to the telephone network by attachment to the handset of a telephone.

APPC   Advanced Program to Program Communications. A set of IBM standards that enables common data to be processed by a variety of computers within an SNA environment.

ARPANET   A packet switching network operated by the U.S. government.

ASCII   American Standard Code for Information Interchange. A set of numbers from zero through 127 assigned to letters, numbers, punctuation marks, and special characters.

Asynchronous communication   A method of communication in which the intervals between characters are of uneven length.

Baud rate   The length of the shortest signalling condition or event divided into one second.

BISYNC   Binary Synchronous Communications. An IBM synchronous protocol.

Bit   Binary digit. A number that can only have the value 1 or 0.

Block mode   A facility, available only on some terminals, whereby data on a screen can be edited locally and transmitted as a block when the user has finished, rather than character by character.

Break   A special signal used to interrupt a program.

Buffer   An area of memory containing data waiting to be transmitted, or received data waiting to be processed.

BSC   Another name for BISYNC.

Bulletin board   An electronic database holding messages that can be read and downloaded by a number of users.

B.p.s.   Bits per second. The number of binary digits of information transmitted in one second.

CD   Carrier Detect. Used by a modem to indicate the presence of a carrier signal (line 8 on RS-232 connections).

Coaxial cable   A cable consisting of a central wire surrounded by another wire in the form of a cylinder.

CompuServe   An organization that offers a number of services to its subscribers including electronic mail, database, and conference facilities.

Computer Conferencing   A facility whereby several users can talk to each other simultaneously by sending messages to a central computer that sends the information on to all the users currently on-line.

CRC   Cyclical Redundancy Check. An error checking method of computing a number from a string of transmitted data that is computed by both the transmitting and receiving device.

CTS   Clear to Send. Secondary handshaking line from DCE to DTE (line 5 in RS-232 connections).

Data bits   The bits forming part of a single group of bits that represent data, as opposed to start, parity, and stop bits.

DCA   Document Content Architecture. A standard data format devised by IBM to enable the same documents to be processed by otherwise incompatible computers.

DCE   Data Communication Equipment. Modems and other intermediate communications devices (distinguished from DTE).

DIALOG   An organization that provides access to a large number of databases at fees that vary according to the database used.

Direct connection   Used in this book to refer to devices connected through RS-232 wires rather than through modems.

DSR   Data Set Ready. Primary handshaking line from DCE to DTE, (line 6 in RS-232 connections).

DTE   Data Terminal Equipment. Terminals and other devices that are the source or final destination of data, as opposed to DCE devices, which are intermediate communications devices.

DTR   Data Terminal Ready. Primary handshaking line from DTE to DCE (line 20 in RS-232 connections).

EBCDIC   Extended Binary Coded Decimal Interchange Code. An alternative to the ASCII code used mainly by IBM computers other than the IBM PC series.

ENQ   Enquiry, ASCII code 5. Used to ask for ACK or NAK.

ENQ/ACK   A handshaking protocol using the ENQ, ACK, and NAK characters.

Even parity   Adding a bit after the data bits to make the total number of binary ones in the data bits and the parity bit an even number (*see also* Parity check).

Forms caching   A facility on some terminals whereby several screens of data can be entered locally, and subsequently transmitted as one batch.

FSK   Frequency Shift Keying. Used by some modems to encode data as different frequencies.

Full duplex   Simultaneous two-way communication.

Half duplex   Communication in two directions but not at the same time.

Hardware interrupt   A signal sent from a device indicating the occurrence of an event.

HDLC   High Level Data Link Control. A synchronous protocol.

Interrupt   *See* Hardware interrupt, Software interrupt.

ISDN   Integrated Services Digital Network. A system being developed that integrates voice, data, and other communications over common channels.

I/O   Input/Output.

Kermit   A protocol designed for micro to mainframe file transfers. (Also a talking frog.)

LAN   Local Area Network. A system connecting a number of communications devices in one location.

LRC   Longitudinal Redundancy Check. An alternative check to the CRC.

MARK   The communication of a binary 1, represented by a negative voltage in direct connection.

Modem   Modulator Demodulator. A device for converting computer communications to and from a form appropriate for the telephone network.

Multiplex   A method whereby several devices can share a communications channel.

Multipoint   A system where several devices are sharing a single communications line.

NAK   Negative Acknowledge, ASCII code 21 decimal. Used to indicate errors in a received transmission.

Network   A system connecting a number of communications devices.

Node   A part of a network connected to a number of circuits that are consolidated for onward transmission.

Null modem   A set of circuits that enables two DTE or two DCE devices to be connected by swapping the necessary wires.

Odd parity   Adding a bit after the data bits so as to make the total number of binary ones in the data bits and the parity bit an odd number (*see also* Parity check).

Packet   A group of data elements transmitted together that generally forms part of a larger transmission made up of a number of packets. A packet is made up of additional information such as packet number and error detecting codes.

Packet switching   A method of communication that involves splitting a transmission up into packets. Successive packets along a given channel can belong to different transmissions.

PAD   Packet Assembler/Disassembler. A device used to create and unpack packets in packet switching.

PAM   Phase Amplitude Modulation. Used by some modems to translate bits into a combination of phase shifts and frequency changes.

Parity bit   A bit sent after the data bits used for error detection. It is computed from the data bits by both the transmitting and receiving devices and the result is then compared.

Parity error   The condition that arises when the parity bit does not bear the correct relationship to the data bits.

Phase modulation   Used by some modems to translate bits into different phases of a carrier signal.

PIC   Programmable Interrupt Controller. A chip used to enable and prioritize hardware interrupts and pass them on to the CPU.

Point to point   The opposite of multipoint. A communications line is being used by only one device at each end.

Polling   A CPU repeatedly examines a number of devices in turn to see whether anything has happened as opposed to interrupt-driven communications where the devices themselves notify the CPU when something happens.

Protocol   A set of standards covering data communications.

PSK   Phase Shift Keying. Used by some modems to encode data as different phase angles.

RI   Ring Indicator. Used by a modem to indicate that it is receiving a call and would be ringing if it were a telephone. (line 22 in RS-232 connections).

RQS or RTS   Request to Send. Secondary handshaking line from DTE to DCE (line 4 in RS-232 connections).

RXD   Received Data. The circuit carrying data from DCE to DTE (line 3 in RS-232 connections).

SDLC   Synchronous Data Link Control. A bit-oriented synchronous protocol.

Serial communications   The transmission of data as a sequence of bits.

SG   Signal Ground. A common reference point for various circuits (line 7 in RS-232 communications).

SNA　Systems Network Architecture. A method used by IBM for connecting computers in a network.

Software interrupt　Software use of a computer's interrupt mechanism to cause the execution of a section of code associated with a particular interrupt number.

SPACE　The communication of a binary zero, represented in direct connection by a positive voltage.

Start bit　A bit sent in asynchronous communications indicating the start of a new character.

Stop bit　A bit sent in asynchronous communications indicating the end of a character.

Synchronous communications　The transmission of a sequence of data elements at regularly spaced intervals without the use of start and stop bits framing each character.

VRC　Vertical Redundancy Check. The use of odd or even parity as an error-checking mechanism.

TDM　Time Division Multiplexing. The use of a single line by several devices, each with its own time slot.

Telenet　A commercial packet-switching network.

Timeout　A period after which, if no response is received, an error is considered to have occurred.

TXD　Transmitted Data. The line carrying data from DTE to DCE (line 2 in RS-232 connections).

Tymnet　A commercial packet-switching network.

UART　Universal Asynchronous Receiver/Transmitter. A chip with serial/parallel conversion, parallel/serial conversion, and other facilities designed for use in asynchronous serial communications.

USART   Universal Synchronous/Asynchronous Receiver/Transmitter. It is similar to a UART, but has synchronous capability also.

XMODEM   A protocol designed for transfers between microcomputers.

X.25   A protocol used by the packet switching networks.

X.PC   An asynchronous version of X.25.

YMODEM   An enhanced version of XMODEM.

WATS   Wide Area Telephone Service. Unlimited use of a telephone circuit for specified periods for an agreed charge.

Word length   The number of data bits sent at one time during asynchronous communications.

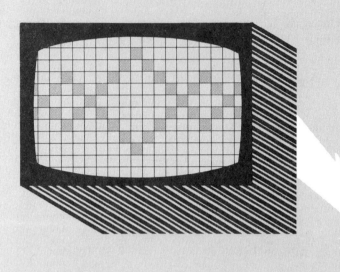

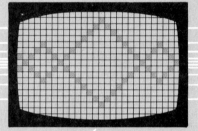

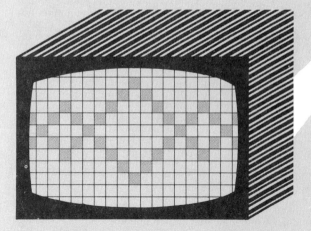

APPENDIX **C**

Bibliography

BIBLIOGRAPHY

Campbell, Joe. *The RS-232 Solution.* Berkeley: SYBEX, Inc., 1984

da Cruz, Frank. "Kermit Protocol Manual." Photocopied. New York: Columbia University Center for Computing Activities, 1984.

Eggebrecht, Lewis C. *Interfacing to the IBM Personal Computer.* Indianapolis: Howard Sams, 1983.

Electronic Industries Association. "EIA-232-C." Photocopied. Washington D.C.: Electronic Industries Association, 1981.

———. "EIA-422-A." Photocopied. Washington D.C.: Electronic Industries Association, 1978.

———. "Industrial Electronics Bulletin No. 12—Application Notes on Interconnection Between Interface Circuits Using RS-449 and RS-232-C." Photocopied. Washington D.C.: Electronic Industries Association, 1978.

———. "RS-423-A." Photocopied. Washington D.C.: Electronic Industries Association, 1978.

———. "RS-449." Photocopied. Washington D.C.: Electronic Industries Association, 1977.

Glossbrenner, Alfred. *The Complete Handbook of Personal Computer Communications.* New York: St. Martin's Press, 1985

IBM Corporation. *DOS Technical Reference Manual.* Boca Raton, FL: IBM, 1985.

———. *Technical Reference Personal Computer XT.* Boca Raton, FL: IBM, 1983.

King, Richard. *The IBM PC-DOS Handbook.* Berkeley: SYBEX, Inc. 1983.

Liu, Yu-cheng and Gibson, Glenn A. *Microcomputer Systems: The 8086/8088 Family.* Englewood Cliffs, NJ: Prentice Hall, 1986.

McNamara, John E. *Technical Aspects of Data Communication.* Bedford, MA: Digital Press, 1982.

Nichols, Elizabeth A.; Nichols, Joseph C.; and Musson, Keith R. *Data Communications for Microcomputers.* New York: McGraw-Hill, 1982.

Norton, Peter. *Inside the IBM PC.* Bowie, MD: Brady, 1983.

———. *Programmer's Guide to the IBM PC.* Bellevue, WA: Microsoft Press, 1985.

Sherman, Kenneth. *Data Communications.* Reston: Reston Publishing Co., 1985.

INDEX

Selections from The SYBEX Library

Languages

PASCAL

INTRODUCTION TO TURBO PASCAL
by Douglas S. Stivison
268 pp., illustr., Ref. 269-8
This bestseller introduces Pascal programming in the environment of Turbo Pascal, giving realistic examples from the author's programming experience. The focus is on how to get all the benefits offered by this Pascal implementation.

INTRODUCTION TO PASCAL, INCLUDING TURBO PASCAL
by Rodnay Zaks
464 pp., illustr., Ref. 319-8
This new version of the Sybex classic book describes Pascal clearly and quickly. There is a complete set of exercises and answers in both Turbo Pascal and ISO Standard Pascal.

TURBO PASCAL LIBRARY
by Douglas S. Stivison
221 pp., illustr., Ref. 330-9
This presents a collection of proven programs and procedures that express Turbo's style and power. The library includes general-purpose procedures applicable to a wide range of programming projects including games, system utilities, and calculating routines for business and engineering applications. Ideal for students, new programmers, and experienced programmers looking to increase their Turbo resources.

INTRODUCTION TO PASCAL (Including UCSD Pascal)
by Rodnay Zaks
420 pp., 130 illustr., Ref. 066-0

A step-by-step introduction for anyone who wants to learn the Pascal language, describing UCSD and Standard Pascals. No technical background is assumed.

THE PASCAL HANDBOOK
by Jacques Tiberghien
486 pp., 270 illustr., Ref. 053-9
A dictionary of the Pascal language, defining every reserved word, operator, procedure, and function found in all major versions of Pascal.

PASCAL PROGRAMS FOR SCIENTISTS AND ENGINEERS
by Alan R. Miller
374 pp., 120 illustr., Ref. 058-X
A comprehensive collection of frequently used algorithms for scientific and technical applications, programmed in Pascal. Includes programs for curve-fitting, integrals, stastical techniques, and more.

FIFTY PASCAL PROGRAMS
by Bruce H. Hunter
338 pp., illustr., Ref. 110-1
More than just a collection of useful programs! Structured programming techniques are emphasized and concepts such as data type creation and array manipulation are clearly illustrated.

THE C LANGUAGE

UNDERSTANDING C
by Bruce H. Hunter
320 pp., Ref. 123-3
Explains how to program in powerful C language for a variety of applications. Some programming experience assumed.

MASTERING C
by Craig Bolon
400 pp., illustr., Ref. 326-0

Designed for the programming professional, this gives a complete description of C language programming, focusing on how to get the most power, efficiency, and portability out of C.

DATA HANDLING UTILITIES IN C
**by Robert Radcliffe
and Thomas Raab**
500 pp., illustr., Ref. 304-X
This is a "Software Toolkit" of useful C functions, techniques and usable code for commercial application programmers and software developers. Because commercial programs require high user-interaction and permanent files, the book concentrates on data entry, validation, display, and efficient data storage. There is a comprehensive section all about logical data types and another giving sample applications.

Technical

ASSEMBLY LANGUAGE

ASSEMBLY LANGUAGE TECHNIQUES FOR THE IBM PC
by Alan Miller
350 pp., illustr., Ref. 309-0
Any IBM PC user and programmer that wants to learn techniques to get more power from the PC will find the tutorial and program library elements in this title extremely valuable. Programs included in the book allow the reader to do such tasks as transferring WordStar to ASCII and back, switch from color screens to monochrome screens and back, set the printer to any typeface, and more. Techniques are given for the programmer to generate more programs.

PROGRAMMING THE 65816
by William Labiak
350 pp., illustr., Ref. 324-4
Giving the latest in this hot new area of development, this book teaches assembly language programming for the 65816, 65C816, 65S816, and 65SC816 chips. The 65802 is also presented. Step-by-step exercises and tutorials enable the reader to write complete applications programs.

PROGRAMMING THE APPLE II IN ASSEMBLY LANGUAGE
by Rodnay Zaks
519 pp., illustr., Ref. 290-6
All elements of the art of assembly language programming for the current Apple IIc and Apple IIe are covered in Zaks' classic style.

PROGRAMMING THE MACINTOSH IN ASSEMBLY LANGUAGE
by Steve Williams
400 pp., illustr., Ref. 263-9
This is an up-to-date tutorial and reference guide to programming the 68000 in the Macintosh environment. Covering architecture, instruction set, Toolbox, and advanced programming concepts, this is ideal for intermediate to professional applications programmers.

HARDWARE

MICROPROCESSOR INTERFACING TECHNIQUES
by Rodnay Zaks and Austin Lesea
456 pp., 400 illustr., Ref. 029-6
Complete hardware and software interfacing techniques, including D to A conversion, peripherals, bus standards and troubleshooting.

THE RS-232 SOLUTION
by Joe Campbell
194 pp., illustr., Ref. 140-3
Finally, a book that will show you how to correctly interface your computer to any RS-232-C peripheral.

MASTERING SERIAL COMMUNICATIONS
by Joe Campbell
250 pp., illustr., Ref. 180-2
This sequel to "The RS-232 Solution" guides the reader to mastery of more complex interfacing techniques.

OPERATING SYSTEMS

REAL WORLD UNIX
by John D. Halamka
209 pp., Ref. 093-8

This book is written for the beginning and intermediate UNIX user in a practical, straightforward manner, with specific instructions given for many business applications.

THE PROGRAMMER'S GUIDE TO TOPVIEW
by David K. Simerly
313 pp., illustr., Ref. 273-6
This guides the programmer through all the major features of TopView for the entire IBM PC line. This includes examples of programs on TopView, descriptions of subroutine calls and macros, and instructions for writing including assembly language programming. Special emphasis is given to writing programs that run both with or without TopView.

IBM PC AND COMPATIBLES

OPERATING THE IBM PC NETWORKS
Token Ring and Broadband
by Paul Berry
363 pp., illustr., Ref. 307-4
This tells you how to plan, install, and use either the Token Ring Network or the PC Network. Focusing on the hardware-independent PCN software, this book gives readers who need to plan, set-up, operate, and administrate such networks the head start they need to see their way clearly right from the beginning.

THE ABC'S OF THE IBM PC
by Joan Lasselle and Carol Ramsay
143 pp., illustr., Ref. 102-0
This book will take you through the first crucial steps in learning to use the IBM PC.

THE IBM PC-DOS HANDBOOK
by Richard Allen King
296 pp., Ref. 103-9
Explains the PC disk operating system. Get the most out of your PC by adapting its capabilities to your specific needs.

BUSINESS GRAPHICS FOR THE IBM PC
by Nelson Ford
259 pp., illustr. Ref. 124-1
Ready-to-run programs for creating line graphs, multiple bar graphs, pie charts and more. An ideal way to use your PC's business capabilities!

THE IBM PC CONNECTION
by James Coffron
264 pp., illustr., Ref. 127-6
Teaches elementary interfacing and BASIC programming of the IBM PC for connection to external devices and household appliances.

DATA FILE PROGRAMMING ON YOUR IBM PC
by Alan Simpson
219 pp., illustr., Ref. 146-2
This book provides instructions and examples for managing data files in BASIC Programming. Design and development are extensively discussed.

THE MS-DOS HANDBOOK
by Richard Allen King (2nd Ed)
320 pp., illustr., Ref. 185-3
The differences between the various versions and manufacturer's implementations of MS-DOS are covered in a clear straightforward manner. Tables, maps, and numerous examples make this the most complete book on MS-DOS available.

ESSENTIAL PC-DOS
by Myril and Susan Shaw
300 pp., illustr., Ref. 176-4
Whether you work with the IBM PC, XT, PC jr. or the portable PC, this book will be invaluable both for learning PC DOS and for later reference.

Software Specific

SPREADSHEETS

MASTERING SUPERCALC 3
by Greg Harvey
300 pp., illustr., Ref. 312-0

Featuring Version 2.1, this title offers full coverage of all the sophisticated features of this third generation spreadsheet, including spreadsheet, graphics, database and advanced techniques.

DOING BUSINESS WITH MULTIPLAN
by Richard Allen King and Stanley R. Trost
250 pp., illustr., Ref. 148-9
This book will show you how using Multiplan can be nearly as easy as learning to use a pocket calculator. It presents a collection of templates for business applications.

MULTIPLAN ON THE COMMODORE 64
by Richard Allen King
250 pp., illustr. Ref. 231-0
This clear, straightforward guide will give you a firm grasp on Multiplan's function, as well as provide a collection of useful template programs.

WORD PROCESSING

PRACTICAL WORDSTAR USES
by Julie Anne Arca
303 pp., illustr. Ref. 107-1
Pick your most time-consuming office tasks and this book will show you how to streamline them with WordStar.

THE COMPLETE GUIDE TO MULTIMATE
by Carol Holcomb Dreger
250 pp., illustr. Ref. 229-9
A concise introduction to the many practical applications of this powerful word processing program.

THE THINKTANK BOOK
by Jonathan Kamin
200 pp., illustr., Ref. 224-8
Learn how the ThinkTank program can help you organize your thoughts, plans and activities.

PRACTICAL MULTIMATE USES
by Chris Gilbert
275 pp., illustr., Ref. 276-0

Includes an overview followed by practical business techniques, this covers documentation, formatting, tables, and Key Procedures.

MASTERING WORDSTAR ON THE IBM PC
by Arthur Naiman
200 pp., illustr., Ref. 250-7
The classic Introduction to WordStar is now specially presented for the IBM PC, complete with margin-flagged keys and other valuable quick-reference tools.

MASTERING MS WORD
by Mathew Holtz
365 pp., illustr., Ref. 285-X
This clearly-written guide to MS WORD begins by teaching fundamentals quickly and then putting them to use right away. Covers material useful to new and experienced word processors.

PRACTICAL TECHNIQUES IN MS WORD
by Alan R. Neibauer
300 pp., illustr., Ref. 316-3
This book expands into the full power of MS WORD, stressing techniques and procedures to streamline document preparation, including specialized uses such as financial documents and even graphics.

INTRODUCTION TO WORDSTAR 2000
by David Kolodney and Thomas Blackadar
292 pp., illustr., Ref. 270-1
This book covers all the essential features of WordStar 2000 for both beginners and former WordStar users.

PRACTICAL TECHNIQUES IN WORDSTAR 2000
by John Donovan
250 pp., illustr., Ref. 272-8
Featuring WordStar 2000 Release 2, this book presents task-oriented tutorials that get to the heart of practical business solutions.

MASTERING THINKTANK ON THE 512K MACINTOSH
by Jonathan Kamin
264 pp., illustr., Ref. 305-8
Idea-processing at your fingertips: from basic to advanced applications, including answers to the technical question most frequently asked by users.

DATABASE MANAGEMENT SYSTEMS

UNDERSTANDING dBASE III PLUS
by Alan Simpson
415 pp., illustr., Ref. 349-X
Emphasizing the new PLUS features, this extensive volume gives the database terminology, program management, techniques, and applications. There are hints on file-handling, debugging, avoiding syntax errors.

UNDERSTANDING dBASE III
by Alan Simpson
250 pp., illustr., Ref. 267-1
The basics and more, for beginners and intermediate users of dBASEIII. This presents mailing label systems, bookkeeping and data management at your fingertips.

ADVANCED TECHNIQUES IN dBASE III
by Alan Simpson
505 pp., illustr., Ref. 282-5
Intermediate to experienced users are given the best database design techniques, the primary focus being the development of user-friendly, customized programs.

MASTERING dBASE III: A STRUCTURED APPROACH
by Carl Townsend
338 pp., illustr., Ref. 301-5
Emphasized throughout is the highly successful structured design technique for constructing reliable and flexible applications, from getting started to advanced

techniques. A general ledger program is used as the primary illustration for the examples.

UNDERSTANDING dBASE II
by Alan Simpson
260 pp., illustr., Ref. 147-0
Learn programming techniques for mailing label systems, bookkeeping, and data management, as well as ways to interface dBASE II with other software systems.

ADVANCED TECHNIQUES IN dBASE II
by Alan Simpson
395 pp., illustr. Ref., 228-0
Learn to use dBASE II for accounts receivable, recording business income and expenses, keeping personal records and mailing lists, and much more.

INTEGRATED SOFTWARE

MASTERING 1-2-3
by Carolyn Jorgensen
420 pp., illustr., Ref. 337-6
This book goes way beyond using 1-2-3, adding powerful business examples and tutorials to thorough explanations of the program's complex features. Detailing multiple functions, powerful commands, graphics and database capabilities, macros, and add-on product support from Report Writer, Spotlight, and The Cambridge Spread-sheet Analyst. Includes Release 2.

SIMPSON'S 1-2-3 MACRO LIBRARY
by Alan Simpson
300 pp., illustr., Ref. 314-7
This book provides many programming techniques, macro examples, and entire menu-driven systems that demonstrate the full potential of macros. The full power of 1-2-3 version 2 is laid out in powerful, time-saving business solutions developed by bestselling author Alan Simpson.

Introduction to Computers

THE SYBEX PERSONAL COMPUTER DICTIONARY

120 pp. Ref. 199-3

All the definitions and acronyms of micro computer jargon defined in a handy pocket-sized edition. Includes translations of the most popular terms into ten languages.

FROM CHIPS TO SYSTEMS: AN INTRODUCTION TO MICROPROCESSORS

by Rodnay Zaks

552 pp., 400 illustr., Ref. 063-6

A simple and comprehensive introduction to microprocessors from both a hardware and software standpoint: what they are, how they operate, how to assemble them into a complete system.

Special Interest

CELESTIAL BASIC

by Eric Burgess

300 pp. 65 illustr. Ref. 087-3

A collection of BASIC programs that rapidly complete the chores of typical astronomical computations. It's like having a planetarium in your own home! Displays apparent movement of stars, planets and meteor showers.

PERSONAL COMPUTERS AND SPECIAL NEEDS

by Frank G. Bowe

175 pp., illustr. Ref. 193-4

Learn how people are overcoming problems with hearing, vision, mobility, and learning, through the use of computer technology.

Computer Specific

AMIGA

PROGRAMMER'S REFERENCE GUIDE TO THE AMIGA

by Eugene Mortimore

530 pp., illustr., Ref. 343-0

Here is the singlular reference that puts the Amiga's power at a programmer's fingertips. The detailed compendium of Amiga system facilities gives all the facts needed by software developers, working programmers, and in-depth Amiga users, in A-Z order. There are sections on the ROM-BIO exec calls, Graphics Library, Animation Library, Layers Library, Intuition calls, and the Workbench.

APPLE II - MACINTOSH

THE PRO-DOS HANDBOOK

by Timothy Rice and Karen Rice

225 pp., illustr. Ref. 230-2

All Pro-DOS users, from beginning to advanced, will find this book packed with vital information. The book covers the basics, and then addresses itself to the Apple II user who needs to interface with Pro-DOS when programming in BASIC. Learn how Pro-DOS uses memory, and how it handles text files, binary files, graphics and sound. Includes a chapter on machine language programming.

PROGRAMMING THE MACINTOSH IN ASSEMBLY LANGUAGE

by Steve Williams

400 pp., illustr. Ref. 263-9

Information, examples, and guidelines for programming the 68000 microprocessor are given, including details of its entire instruction set.

USING THE MACINTOSH TOOLBOX WITH C

by Fred A. Huxham, David Burnard and Jim Takatsuka

559 pp., illustr., Ref. 249-3

In one place, all you need to get applications runnning on the Macintosh, given clearly, completely, and understandably. Featuring C.

MASTERING Pro-DOS
by Timothy Rice and Karen Rice
250 pp., illustr., Ref. 315-5
This companion volume to The ProDOS Handbook contains numerous examples of programming techniques and utilities that will be valuable to intermediate and advanced users.

THE EASY GUIDE TO YOUR MACINTOSH
By Joseph Caggiano
214 pp., illustr., Ref. 216-7
Simple and quick to use, this tells first time users how to set up their Macintosh computers and how to use the major features and software.

MACINTOSH FOR COLLEGE STUDENTS
by Bryan Pfaffenberger
250 pp., illustr., Ref. 227-2
Find out how to give yourself an edge in the race to get papers in on time and prepare for exams. This book covers everything you need to know about how to use the Macintosh for college study.

CP/M SYSTEMS

THE CP/M HANDBOOK
by Rodnay Zaks
320 pp., illustr., Ref 048-2
An indispensable reference and guide to CP/M – complete in reference form.

MASTERING CP/M
by Alan Miller
398 pp., illustr., Ref. 068-7
For advanced CP/M users or systems programmers who want maximum use of the CP/M operating system: this book takes up where the CP/M Handbook leaves off.

THE CP/M PLUS HANDBOOK
by Alan Miller
250 pp., illustr., Ref. 158-6
This guide is easy for beginners to understand, yet contains valuable information for advanced users of CP/M Plus.

ADVANCED BUSINESS MODELS WITH 1-2-3
by Stanley R. Trost
250 pp., illustr., Ref. 159-4
If you are a business professional using the 1-2-3 software package, you will find the spreadsheet and graphics models provided in this book easy to use "as is" in everyday business situations.

THE ABC'S OF 1-2-3 (New Ed)
by Chris Gilbert and Laurie Williams
225 pp., illustr., Ref. 168-3
For those new to the LOTUS 1-2-3 program, this book offers step-by-step instructions in mastering its spreadsheet, data base, and graphing capabilities. Features Version 2.

MASTERING SYMPHONY
by Douglas Cobb (2nd Ed)
763 pp., illustr., Ref. 224-8
This bestselling book has been heralded as the Symphony bible, and provides all the information you will need to put Symphony to work for you right away. Packed with practical models for the business user. Includes Version 1.1.

ANDERSEN'S SYMPHONY TIPS & TRICKS
by Dick Andersen and Janet McBeen
325 pp., illustr. Ref. 342-2
Organized as a reference tool, this book gives shortcuts for using Symphony commands and functions, with troubleshooting advice.

BETTER SYMPHONY SPREADSHEETS
by Carl Townsend
287 pp., illustr., Ref. 339-2

For Symphony users who want to gain real expertise in the use of the spreadsheet features, this has hundreds of tips and techniques. There are also instructions on how to implement some of the special features of Excel on Symphony.

MASTERING FRAMEWORK
by Doug Hergert
450 pp., illustr. Ref. 248-5
This tutorial guides the beginning user through all the functions and features of this integrated software package, geared to the business environment.

ADVANCED TECHNIQUES IN FRAMEWORK
by Alan Simpson
250 pp., illustr. Ref. 257-4
In order to begin customizing your own models with Framework, you'll need a thorough knowledge of Fred programming language, and this book provides this information in a complete, well-organized form.

MASTERING THE IBM ASSISTANT SERIES
by Jeff Lea and Ted Leonsis
249 pp., illustr., Ref. 284-1
Each section of this book takes the reader through the features, screens, and capabilities of each module of the series. Special emphasis is placed on how the programs work together.

DATA SHARING WITH 1-2-3 AND SYMPHONY: INCLUDING MAINFRAME LINKS
by Dick Andersen
262 pp., illustr., Ref. 283-3
This book focuses on an area of increasing importance to business users: exchanging data between Lotus software and other micro and mainframe software. Special emphasis is given to dBASE II and III.

MASTERING PARADOX
by Alan Simpson
350 pp., illustr., Ref. 334-1
Everyone's introduction to this unique, menu-driven dbms, from essential operations to complex uses including PAL programming techniques. There are valuable real-world illustrations including a complete mailing lists system, and an inventory, sales, and purchasing system with automatic multiple-table updating.

JAZZ ON THE MACINTOSH
by Joseph Caggiano and Michael McCarthy
431 pp., illustr., Ref. 265-5
Each chapter features as an example a business report which is built on throughout the book in the first section of each chapter. Chapters then go on to detail each application and special effects in depth.

MASTERING EXCEL
by Carl Townsend
454 pp., illustr., Ref. 306-6
This hands-on tutorial covers all basic operations of Excel plus in-depth coverage of special features, including extensive coverage of macros.

APPLEWORKS: TIPS & TECHNIQUES
by Robert Ericson
373 pp., illustr., Ref. 303-1
Designed to improve AppleWorks skills, this is a great book that gives utility information illustrated with every-day management examples.

MASTERING APPLEWORKS
by Elna Tymes
201 pp., illustr., Ref. 240-X
This bestseller presents business solutions which are used to introduce AppleWorks and then develop mastery of the program. Includes examples of balance sheet, income statement, inventory control system, cash-flow projection, and accounts receivable summary.

PRACTICAL APPLEWORKS USES
by David K. Simerly
313 pp., illustr., Ref. 274-4
This book covers a breadth of home and business uses, including combined-function applications, complicated tasks, and even a large section on interfacing AppleWorks with the outside world.

SYBEX Computer Books
are different.

Here is why . . .

At SYBEX, each book is designed with you in mind. Every manuscript is carefully selected and supervised by our editors, who are themselves computer experts. We publish the best authors, whose technical expertise is matched by an ability to write clearly and to communicate effectively. Programs are thoroughly tested for accuracy by our technical staff. Our computerized production department goes to great lengths to make sure that each book is well-designed.

In the pursuit of timeliness, SYBEX has achieved many publishing firsts. SYBEX was among the first to integrate personal computers used by authors and staff into the publishing process. SYBEX was the first to publish books on the CP/M operating system, microprocessor interfacing techniques, word processing, and many more topics.

Expertise in computers and dedication to the highest quality product have made SYBEX a world leader in computer book publishing. Translated into fourteen languages, SYBEX books have helped millions of people around the world to get the most from their computers. We hope we have helped you, too.

For a complete catalog of our publications:

SYBEX, Inc. 2021 Challenger Drive, #100, Alameda, CA 94501
Tel: (415) 523-8233/(800) 227-2346   Telex: 336311

Mastering Serial Communications
Programs Available on Disk

If you'd like to use the programs in this book but don't want to type them in yourself, you can send for a disk, in IBM PC format, containing the programs. To obtain this disk, complete the order form and return it along with $20.00. California residents add 6 percent sales tax. Please allow four weeks for delivery. This offer applies to U.S. residents only and is good until December 31, 1988.

Software Licensing, Inc.
394 University Ave.
Palo Alto, CA 94301

Name_____

Address_____

City/State/ZIP_____

Enclosed is my check or money order.
(Make check payable to *Software Licensing, Inc.*)

SYBEX is not affiliated with Software Licensing, Inc. *and assumes no responsibility for any defect in the disk or program.*